# Collins

## AQA GCSE 9

# Physics

Physics

## AQA
### GCSE 9-1

## Workbook

Nathan Goodman

# Revision Tips

## Rethink Revision

Have you ever taken part in a quiz and thought *'I know this!'* but, despite frantically racking your brain, you just couldn't come up with the answer?

It's very frustrating when this happens but, in a fun situation, it doesn't really matter. However, in your GCSE exams, it will be essential that you can recall the relevant information quickly when you need to.

Most students think that revision is about making sure you **know** stuff. Of course, this is important, but it is also about becoming confident that you can **retain** that *stuff* over time and **recall** it quickly when needed.

## Revision That Really Works

Experts have discovered that there are two techniques that help with all of these things and consistently produce better results in exams compared to other revision techniques.

Applying these techniques to your GCSE revision will ensure you get better results in your exams and will have all the relevant knowledge at your fingertips when you start studying for further qualifications, like AS and A Levels, or begin work.

It really isn't rocket science either – you simply need to:

- **test yourself** on each topic as many times as possible
- **leave a gap** between the test sessions.

## Three Essential Revision Tips

1. **Use Your Time Wisely**

   - Allow yourself plenty of time.
   - Try to start revising at least six months before your exams – it's more effective and less stressful.
   - Your revision time is precious so use it wisely – using the techniques described on this page will ensure you revise effectively and efficiently and get the best results.
   - Don't waste time re-reading the same information over and over again – it's time-consuming and not effective!

2. **Make a Plan**

   - Identify all the topics you need to revise.
   - Plan at least five sessions for each topic.
   - One hour should be ample time to test yourself on the key ideas for a topic.
   - Spread out the practice sessions for each topic – the optimum time to leave between each session is about one month but, if this isn't possible, just make the gaps as big as realistically possible.

3. **Test Yourself**

   - Methods for testing yourself include: quizzes, practice questions, flashcards, past papers, explaining a topic to someone else, etc.
   - Don't worry if you get an answer wrong – provided you check what the correct answer is, you are more likely to get the same or similar questions right in future!

Visit our website for more information about the benefits of these techniques and for further guidance on how to plan ahead and make them work for you.

## www.collins.co.uk/collinsGCSErevision

# Contents

# Forces – An Introduction

1. When a plane is in flight, the engines provide a thrust force that pushes the aircraft forwards. The wings provide a 'lift' force that acts upwards.

    a) Name **two** other forces that act on the plane.

    In each case state whether it is contact force or non-contact force.

    ................................................................................................................................................

    ................................................................................................................................................ [4]

    b) The plane has a mass of 120 000kg. Calculate the weight of the plane ($g$ = 10N/kg).

    Weight = ............................................... N  [2]

    c) The plane accelerates and ascends to a higher altitude.
    During this time, the resultant forwards force is twice the size of the resultant upwards force.

    Use this information to draw a scale
    vector diagram.
    Your diagram should show:
    - The resultant forwards force.
    - The resultant upwards force.
    - A final resultant force that shows the
      combined effect of all the forces acting
      on the plane. [3]

2. A student carries out an investigation into forces.
    They use an air track, which suspends a glider vehicle on a cushion of air so that it can move smoothly without touching the ground.

    a) What force is the air track designed to reduce?

    Answer ............................................... [1]

    b) Use the idea of contact and non-contact forces to explain why the air track is effective at doing this.

    ................................................................................................................................................

    ................................................................................................................................................ [2]

    Total Marks ............... / 12

# Forces in Action

**1**  A student carries out an investigation involving springs.
The student suspends a spring from a rod.
A force is applied to the spring by suspending a mass from it.

a) Describe how the student could test if the spring is behaving elastically.

............................................................................................................................................. [2]

............................................................................................................................................

b) In a second investigation, the student takes a set of measurements for force and extension.
The results are shown in **Table 1** below.

**Table 1**

| Force (N) | 0.0 | 1.0 | 2.0 | 3.0 | 4.0 | 5.0 | 6.0 |
|---|---|---|---|---|---|---|---|
| Extension (cm) | 0.0 | 4.0 | | 12.0 | 16.0 | 22.0 | 31.0 |

   i)  Add the missing value to the table. [1]

   ii) Explain why you chose this value.

............................................................................................................................................

............................................................................................................................................. [2]

**2**  A man uses a long lever to remove tree stump from the ground.
The total length of the lever is 1.8m and the lever is pivoted 20cm from the end.

**Figure 1**

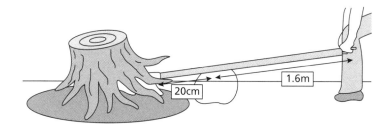

20cm   1.6m

a) The man pushes down with a force of 200N.

   Work out the turning moment created.

   Answer ............................. [2]

b) When pressing down with a force of 200N the man is just able to lift the tree stump.

   Use this information and your answer to part **a)** to work out the force needed to lift the stump.

   Answer ............................. [2]

Total Marks ................. / 9

# Pressure and Pressure Differences

**1** A hydraulic jack uses pressure to transfer a force and lift a car.

**Figure 1** shows the construction of the jack.

The pressure in the liquid is the same at all points.

**Figure 1**

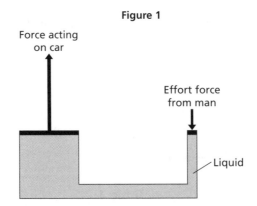

a) Give the equation that links pressure, force and area.

Answer .................................................. [1]

b) Use **Figure 1** to explain how the man using the jack is able to exert a much greater force on the car than he would be able to without using the jack.

......................................................................................................................................................

......................................................................................................................................................

......................................................................................................................................................

[3]

**2** A hot air balloon experiences a force of upthrust caused by the surrounding air.

This force causes the hot air balloon to rise.

a) What can be said about the relative density of the hot air balloon and the air around it?

......................................................................................................................................................

......................................................................................................................................................

[1]

b) The hot air balloon rises until it reaches a steady height.

Use your understanding of forces and the atmosphere to explain why this happens.

......................................................................................................................................................

......................................................................................................................................................

......................................................................................................................................................

......................................................................................................................................................

[3]

**Total Marks** .................... / 8

# Forces and Motion

**1** A person takes their dog for a walk.

The graph in **Figure 1** shows how the distance from their home changes with time.

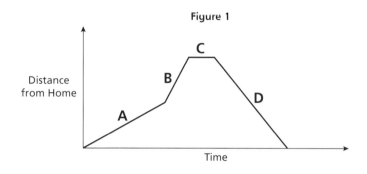

Figure 1

a) Work out the final displacement at the end of their journey.

   Answer ........................................... [1]

b) Which part of the graph, **A**, **B**, **C** or **D**, shows them walking at the fastest rate?

   Answer ........................................... [1]

c) Describe their motion during section **C**.       Answer ........................................... [1]

d) Describe how the velocity of section **A** compares to the velocity of section **D**.

   ..................................................................................................................................................

   ..................................................................................................................................................

   .................................................................................................................................... [3]

**2** Drivers on a racetrack enter a hairpin bend travelling east.

The tight bend forces them to slow down.

When they exit the bend, they are travelling west and speed back up again.

The bend is 180m long.

a) If it takes 6 seconds to travel around the bend, what is the average speed of the car around the bend?

   Speed = ........................................... m/s [2]

b) A driver enters the bend at 50m/s and exits the bend at 40m/s.

   i) Work out the change in speed.

      Speed = ........................................... m/s [1]

   ii) Work out the change in velocity.

      Speed = ........................................... m/s [1]

Total Marks ........................ / 10

# Forces and Acceleration

**1** An experiment is carried out to investigate how changing the mass affects the acceleration of a system.
A trolley is placed on a bench and is made to accelerate by applying a constant force using hanging masses. Different masses were then added to the trolley.

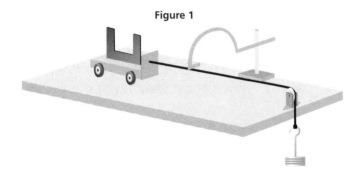

**Figure 1**

a) What is the independent variable?

Answer ........................................................ [1]

b) What is the control variable?

Answer ........................................................ [1]

c) What is the dependent variable?

Answer ........................................................ [1]

**2** A boat accelerates at a constant rate in a straight line.
This causes the velocity to increase from 4.0m/s to 16.0m/s in 8.0s.

a) Calculate the acceleration.
Give the unit.

Answer ........................................................ [3]

b) A water skier being pulled by the boat has a mass of 68kg.

Use your answer to part **a)** to calculate the resultant force acting on the water skier whilst accelerating.

Answer ........................................................ [2]

**3** The manufacturer of a car gives the following information in a brochure:
The mass of the car is 950kg.
The car will accelerate from 0 to 33m/s in 11s.

a) Calculate the acceleration of the car during the 11s.

Answer ........................................................ [2]

b) Calculate the force needed to produce this acceleration.

Answer ........................................................ [2]

**Total Marks** ................ / 12

# Terminal Velocity and Momentum

**1** The terminal velocity is the maximum velocity a falling object can reach.

Describe how a skydiver could increase their terminal velocity.

[2]

**2** A Saturn 5 rocket, as used on the Apollo space missions, has F1 rocket engines.
Each engine burns 5000kg of fuel every second, producing a thrust of 6.7MN.

a) Given that **force = rate of change of momentum**, calculate the speed with which the exhaust gases exit each engine.

Answer _____ m/s [3]

b) During the Apollo mission, five engines were used simultaneously.

Use your answer to part **a)** to calculate the momentum gained by the rocket in the first 10 seconds of travel (assuming there were no other resistive forces).

Answer _____ kg m/s [3]

c) The entire launch vehicle has a mass of 3000 000kg.

Use your answer to part **b)** to calculate the velocity after the first 10 seconds (assuming all other factors are unchanged).

Answer _____ m/s [3]

d) The rocket engines provide constant thrust.
However, after 100 seconds, the acceleration of the rocket is greater than the initial acceleration.

Give **two** reasons why this might be.
You must explain your answers.

[4]

Total Marks _____ / 15

# Stopping and Braking

1  The graph in **Figure 1** shows how the velocity of a car changes from the moment the driver sees an obstacle blocking the road.

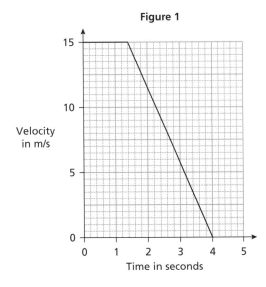

**Figure 1**

a)  Work out the reaction time of the driver.

Answer ........................................................ [1]

b)  Use the graph and your answer to part **a)** to calculate the thinking distance.

Answer ........................................................ [2]

c)  Use the graph to work out the time it took for the car to stop from the moment the brakes were applied.

Answer ........................................................ [1]

d)  The car and driver have a combined mass of 1300kg.

Work out the momentum of the car just before the brakes were applied.

Answer ........................................................ [2]

e)  Use your answers to parts **c)** and **d)** to work out the braking force applied by the brakes of the car.

Answer ........................................................ [3]

f)  The driver of the car was tired and had been drinking alcohol.

On the graph in **Figure 1**, sketch a second line to show how the graph would be different if the driver had been wide awake and fully alert. [3]

g)  How would the graph look different if the vehicle had old / worn brakes and tyres?

.................................................................................................................................

................................................................................................................................. [2]

Total Marks ................ / 14

# Energy Stores and Transfers

**1** An electric kettle is used to heat 2kg of water from 20°C to 100°C.

> change in thermal energy = mass × specific heat capacity × temperature change
>
> Specific heat capacity of water = 4200J/kg°C

**a)** All of the energy supplied to the kettle goes into the water.

Calculate the amount of electrical energy supplied to the kettle.

Answer .................................................... [3]

**b)** On a different occasion, the kettle is filled with 2.5kg of water but is switched on for the same amount of time.

Use your answer to part **a)** to calculate what temperature the kettle heats the water to in this period.

Answer .................................................... [3]

**2** **Figure 1** shows a pendulum swinging backwards and forwards.
The pendulum is made from a 100g mass suspended by a light string.

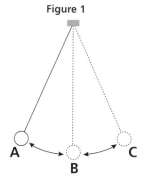

Figure 1

**a)** At position **C**, the mass is 5cm higher than at position **B**.

Calculate the difference in gravitational energy between positions **B** and **C**.
The gravitational field strength is = 10N/kg.

Answer .................................................... [3]

**b)** The difference in potential energy is the same as the amount of kinetic energy gained by the mass as it swings down from **C** to **B**.

Calculate the velocity of the mass at position **B**.

Answer .................................................... [3]

**Total Marks** ............... / 12

# Energy Transfers and Resources

**1** Complete the sentences below to explain the energy transfers involved in a solar panel.

In a solar panel, _____ energy is converted into _____ energy.

Some energy is converted into _____ energy and lost to the surroundings. **[3]**

**2** A student tested four different types of fleece, **J**, **K**, **L** and **M**, to find out which would make the warmest jacket.

Each type of fleece was wrapped around a can. The can was then filled with hot water.
The temperature of the water was recorded every 2 minutes for a 20-minute period.
The graph in **Figure 2** shows the student's results.

**Figure 1**

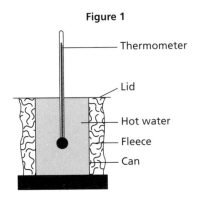

- Thermometer
- Lid
- Hot water
- Fleece
- Can

**Figure 2**

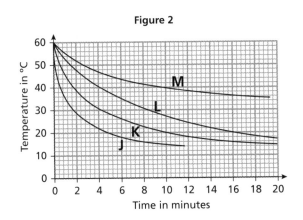

**a)** To be able to compare the results, it was important to use the same volume of water in each test.

Give **two** other variables that should have been kept the same in each test.

_____ **[2]**

**b)** Which type of fleece, **J**, **K**, **L** or **M**, should the student recommend for making a ski jacket?
You must explain your answer.

_____ **[2]**

Total Marks _____ / 7

# Waves and Wave Properties

**1** A wave machine in a swimming pool generates waves with a frequency of 0.5Hz.

a) What does a frequency of 0.5Hz mean?

.................................................................................................................................. [1]

b) Give the equation that links the frequency, speed and wavelength of a wave.

.................................................................................................................................. [1]

c) The swimming pool is 50m long.
It takes each wave 10 seconds to travel the length of the pool.

Calculate the wave speed.

Answer .................................... [2]

d) Use your answers to parts **b)** and **c)** to calculate the wavelength of the waves.

Answer .................................... [2]

e) One section of the swimming pool is designed for young children and has much shallower water.
A parent notices that the waves get closer together when they enter this section.

What effect will this have on the wave speed?

.................................................................................................................................. [1]

**2** Waves may be longitudinal or transverse.

Describe the differences between longitudinal and transverse waves.

..................................................................................................................................

..................................................................................................................................

..................................................................................................................................

..................................................................................................................................

..................................................................................................................................

.................................................................................................................................. [3]

Total Marks .................... / 10

# Reflection, Refraction and Sound

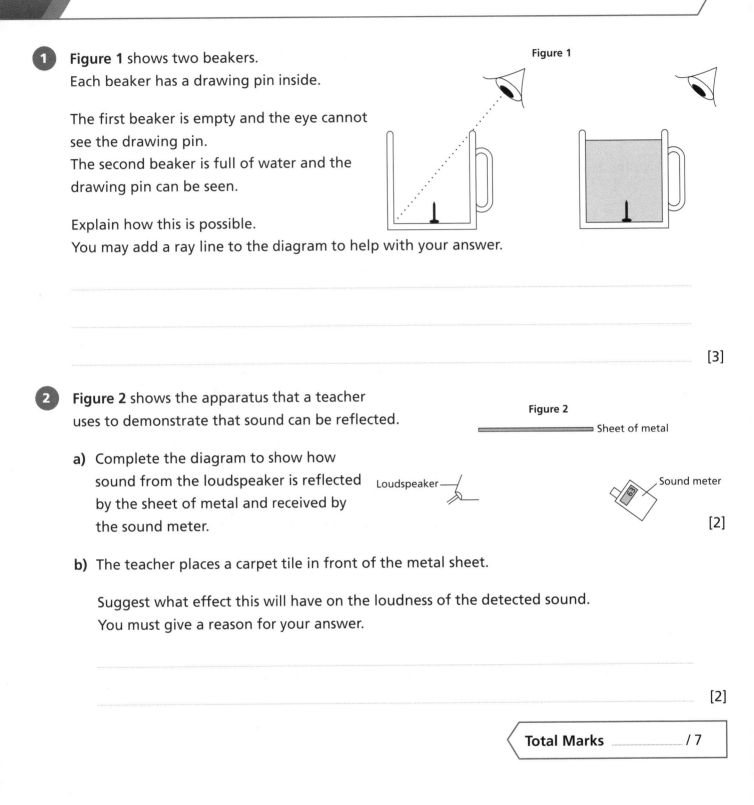

**1** **Figure 1** shows two beakers.
Each beaker has a drawing pin inside.

Figure 1

The first beaker is empty and the eye cannot see the drawing pin.
The second beaker is full of water and the drawing pin can be seen.

Explain how this is possible.
You may add a ray line to the diagram to help with your answer.

..................................................................................................................................

..................................................................................................................................

..................................................................................................................................

[3]

**2** **Figure 2** shows the apparatus that a teacher uses to demonstrate that sound can be reflected.

Figure 2

Sheet of metal

**a)** Complete the diagram to show how sound from the loudspeaker is reflected by the sheet of metal and received by the sound meter.

Loudspeaker

Sound meter

[2]

**b)** The teacher places a carpet tile in front of the metal sheet.

Suggest what effect this will have on the loudness of the detected sound.
You must give a reason for your answer.

..................................................................................................................................

..................................................................................................................................

[2]

Total Marks .................... / 7

# Waves for Detection and Exploration

**1** Ultrasound can be used to measure the depth of water below a ship.
A pulse of ultrasound is emitted by an electronic system on-board the ship.
It takes 0.8 seconds for the ultrasound to be received back at the ship.

Calculate the depth of the water.
Speed of ultrasound in water = 1600m/s

Answer ............................................ [3]

**2** Ultrasound can also be used to detect internal cracks in materials.
An ultrasound pulse is transmitted into the material and any reflected pulses are measured by a detector.
**Figure 1** shows the screen of an oscilloscope connected to a detector, which is being used to examine a steel block.

**Figure 1**

a) Pulse **A** is reflected back by an internal crack.

What is Pulse **B** reflected back from?

Answer ............................................ [1]

b) The metal block is 120mm from front to back.

Study the oscilloscope trace and work out the distance, in millimetres, from the front of the block to the internal crack.

Answer ............................................ [1]

c) When investigating a second steel block, the oscilloscope settings are left exactly the same. Pulse A is located at the same position, but the amplitude of the reflected pulse is 1 square high.

What conclusion can be drawn about the size of the crack in the second steel block?

.................................................................................................................... [1]

Total Marks ............... / 6

# The Electromagnetic Spectrum

1 Radio waves and visible light are electromagnetic waves that are used for communication.

a) Name another type of electromagnetic wave that is used for communication.

Answer ........................................................ [1]

b) Name an electromagnetic wave which is **not** used for communication and give one of its uses.

.................................................................................................................................

.................................................................................................................................. [2]

2 Different parts of the electromagnetic spectrum have different uses.
**Figure 1** shows the electromagnetic spectrum.

**Figure 1**

| Radio waves | Microwaves | Infrared | Visible light | Ultraviolet | X-rays | Gamma rays |
|---|---|---|---|---|---|---|

Complete the sentence below using words from the box.

amplitude    frequency    speed    wavelength

The arrow in the diagram points in the direction of increasing ...........................

and decreasing ...................... . [2]

3 After a person is injured, a doctor will sometimes request a photograph of the patient's bones.

a) Which type of electromagnetic radiation would be used to produce the photograph?

Answer ........................................................ [1]

b) What properties of this radiation enable it to be used to photograph bones?

.................................................................................................................................

.................................................................................................................................

.................................................................................................................................. [2]

Total Marks .................... / 8

# Lenses

**1** The magnification of a lens can be found by dividing the actual height of object by the height of the image.

a) A lens produces a 2cm tall image of an 8cm tall object.

What is the magnification of the lens?          Answer ........................................ [2]

b) A cinema projector produces a 3m tall image from a 2cm tall object.

What is the magnification of the projector lens?          Answer ........................................ [2]

**2** Explain why concave lenses are also called diverging lenses.

.................................................................................................................................................................

.................................................................................................................................................................

.......................................................................................................................................................... [2]

**3** A lens is used to produce an image.
The image produced is real, inverted and magnified.

a) What kind of lens has been used?          Answer ........................................ [1]

b) Where is the image in relation to the lens?
   Tick **one** box.

   Closer to the lens than the object          ☐

   More distant from the lens than the object          ☐

   The same distance from the lens as the object          ☐          [1]

**4** **Figure 1** shows the position of an object formed by a lens.

a) What type of lens is shown in **Figure 1**?

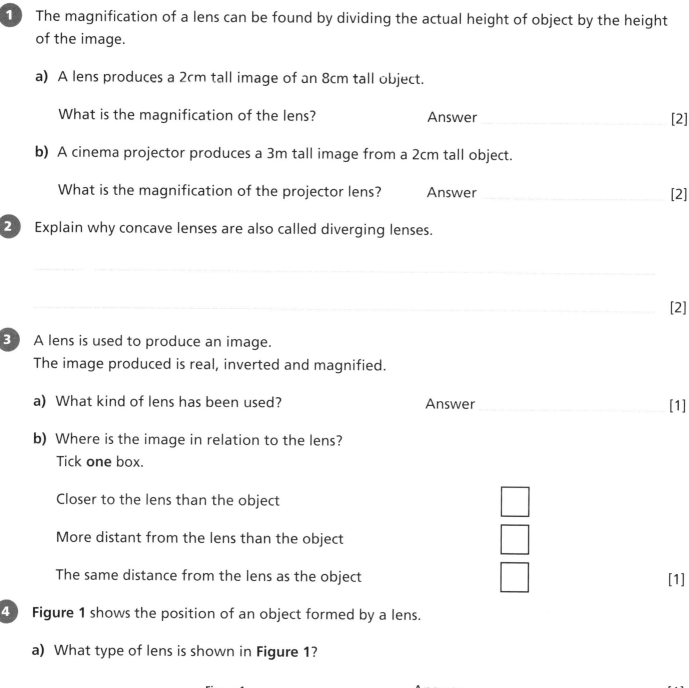

**Figure 1**

Answer ........................................ [1]

b) Construct a ray diagram on **Figure 1** to show how the image is formed.   [4]

c) Use your ray diagram to calculate the magnification of the lens.

Magnification = ........................................ [2]

Total Marks ................. / 15

# Light and Black Body Radiation

1  A trainee lighting technician is trying to mix red and green light.
The trainee places a red and a green filter in front of a white spotlight.
**Figure 1** shows how this has been set up.

Figure 1

White light → Red filter → Green filter → Screen

a) No light reaches the screen.

Explain why.

........................................................................................

........................................................................................

........................................................................................ [2]

b) The trainee lighting technician cuts a circle in the red filter.

What will be seen on the screen now?

........................................................................................ [1]

2  An astronomer is studying distant stars using an infrared telescope.
She finds a brown dwarf star and notes down the position of the star.
She then switches to a visible light telescope, but is not able to see the star.

a) What does this tell the astronomer about the wavelength of light emitted by the star?

........................................................................................ [1]

b) What conclusion can be drawn about the temperature of the star compared to the Sun?

........................................................................................ [1]

c) Use the idea of black body radiation to explain your answers to **a)** and **b)**.

........................................................................................

........................................................................................

........................................................................................

........................................................................................ [3]

**Total Marks** .............. / 8

# An Introduction to Electricity

**1** Circle the correct words to complete the sentences.

Electric **current / charge** is the flow of electrical **charge / potential**.

The **greater / smaller** the flow, the higher the **current / voltage**. [4]

**2** The element in a set of hair straighteners has a 5A current running through it and a 230V potential difference across it.

a) Write down the equation that links potential difference, current and resistance.

Answer ................................................. [1]

b) Calculate the resistance of the element.

Answer ................................................. [2]

c) Write down the equation that links charge, current and time.

Answer ................................................. [1]

d) The straighteners are used for 2 minutes.

Calculate the charge that flows in this time.

Answer ................................................. [2]

**3** Circle the correct words to complete the sentences.

Potential difference determines the amount of **energy / power** transferred by the charge as it passes through a component.

The **greater / smaller** the potential difference, the higher the **current / voltage** that will flow. [3]

**4** Write the name of the component that each circuit symbol represents.

a) Answer ................................................. [1]

b) Answer ................................................. [1]

c) Answer ................................................. [1]

d) Answer ................................................. [1]

Total Marks ................................. / 17

# Circuits and Resistance

**1** The circuit in **Figure 1** is used to measure the current and potential difference of various components.

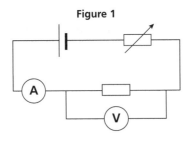

**Figure 1**

a) What is the purpose of the variable resistor?

............................................................................................................................................

............................................................................................................................................ [2]

b) Is the ammeter in **Figure 1** connected in series or parallel with the variable resistor?

Answer ........................................................ [1]

c) Is the voltmeter in **Figure 1** connected in series or parallel with the resistor?

Answer ........................................................ [1]

**2** Draw **one** line from each component to the correct description.

| Light dependent resistor (LDR) | | Resistance decreases as temperature increases. |

| Thermistor | | Resistance increases as the temperature increases. |

| Diode | | Resistance decreases as light intensity increases. |

| Filament light | | Has a very high resistance in one direction. | [3]

Total Marks ................ / 7

# Circuits and Power

1. This question is about a hairdryer that heats air and blows it out the front through a nozzle.

   a) The hairdryer has an input power of 1600W.

   If a person takes two minutes to dry their hair, how much energy has been transferred?

   Answer ........................................... [3]

   b) The hairdryer is 90% efficient. The remaining energy is output as sound.

   How much sound energy is produced in two minutes of use?

   Answer ........................................... [2]

2. **Figure 2** shows a series circuit with two resistors, **X** and **Y**.

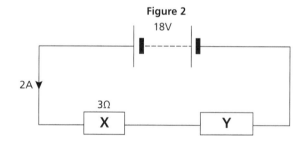

**Figure 2**

   a) Calculate the potential difference across resistor **X**.

   Answer ........................................... [2]

   b) Use your answer to part **a)** to work out the potential difference across component **Y**.

   Answer ........................................... [2]

   c) Calculate the total resistance of the circuit.

   Answer ........................................... [2]

   Total Marks .................... / 11

# Domestic Uses of Electricity

1    A battery is connected to an oscilloscope and the trace in **Figure 1** is produced.

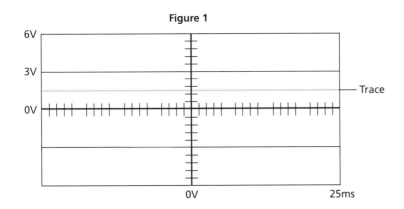

**Figure 1**

a)   Use the trace to determine the potential difference of the battery and the type of current.

                 [2]

b)   The battery is replaced by the mains supply and the trace in **Figure 2** is recorded by the oscilloscope.

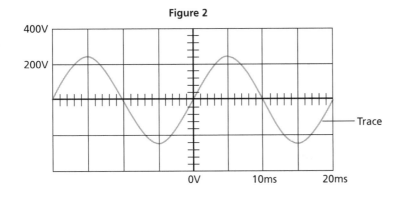

**Figure 2**

Use the trace to determine the potential difference and the type of current.

                 [2]

2    An appliance is switched off but the power cable is connected to the mains.

Explain how a live wire can still be dangerous.

                 [4]

**Total Marks**      / 8

# Electrical Energy in Devices

**1** An electric blender transfers electrical energy into kinetic, heat and sound energy.

a) What is the useful energy output?          Answer _____ [1]

b) What happens to the waste energy produced?

_____

_____ [2]

**2** **Table 1** gives some information about an electric drill.

Table 1

| Energy Input | |
|---|---|
| Useful Energy Output | |
| Wasted Energy | Heat and Sound |
| Power Rating | 500W |

a) Complete **Table 1** by adding the missing types of energy. [2]

b) Which of the following statements about the energy from the drill is **incorrect**?
Tick **one** box.

It spreads out and becomes more difficult to use.   ☐

It disappears.   ☐

It makes the surroundings warmer.   ☐   [1]

c) How much energy does the drill use per second?   Answer _____ [1]

**3** The National Grid distributes electricity from power stations to consumers.
The voltage across the overhead cables of the National Grid is much higher than the output voltage from the power station generators.

Explain how this achieved and why it is important.

_____

_____

_____

_____ [4]

Total Marks _____ / 11

# Static Electricity

**1** Write down whether each of the statements about electric fields around objects is **true** or **false**.

a) The strength of the field depends on the size of the charge on the object.

Answer ........................................ [1]

b) The strength of the field depends on whether the charge is positive or negative.

Answer ........................................ [1]

c) The strength of the field depends on the material that the object is made from.

Answer ........................................ [1]

d) The strength of the field depends on the distance from the object.

Answer ........................................ [1]

**2** a) A student rubs a Perspex rod with a woollen jumper and the Perspex rod becomes positively charged. Describe what has taken place to make this happen.

.......................................................................................................................

.......................................................................................................................

.......................................................................................................................

.......................................................................................................................  [3]

b) **Figure 1** shows what happens when different charged rods interact.

Use the information in **Figure 1** to help you complete the sentence.

**Figure 1**

Perspex rod repels a perspex rod

Perspex rod attracts an ebonite rod

Like charges ........................... and unlike charges ........................... . [2]

**3** Complete the sentences using words from the box.
Each word may be used once, twice or not at all.

| charges | force | electric | magnetic | energy | charge | charged |
|---------|-------|----------|----------|--------|--------|---------|

When a ........................... object is placed in an ........................... field caused by another object,

it will experience a ........................... .

The direction of the ........................... depends on the ........................... of the objects. [5]

Total Marks ........................... / 14

# Magnetism and Electromagnetism

1   The full name for the north pole of a magnet is the 'north-seeking pole'.

Explain what is meant by this.

............................................................................................................................

............................................................................................................................ [2]

2   The north pole of a permanent magnet is moved close to the north pole of another permanent magnet.

a)   What would you expect to happen?

............................................................................................................................ [1]

b)   A piece of iron is moved close to the north pole of a permanent magnet.

What would you expect to happen?

............................................................................................................................ [1]

3   **Figure 1** shows an electric bell.

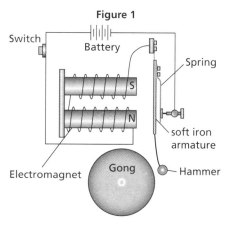

**Figure 1**

Describe what happens when the switch is pressed.

............................................................................................................................

............................................................................................................................

............................................................................................................................

............................................................................................................................ [5]

Total Marks .................... / 9

# The Motor Effect

① **Figure 1** shows a loudspeaker.
Loudspeakers use the motor effect to convert electrical energy into sound energy.

**Figure 1**

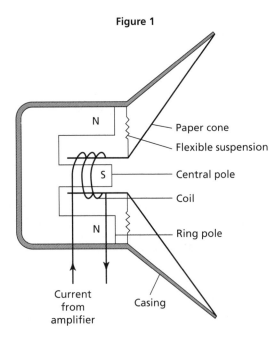

Explain how the input electrical signal is converted into sound.

_____

_____

_____

_____

_____ [4]

② A current carrying wire passes through a magnetic field at right-angles to the field and experiences a force.

The length of wire in the field is 5cm, the current is 2A and the magnetic field is 0.3mT.

a) Calculate the force on the wire.
Use the correct equation from the Physics Equation Sheet on page 71.

Answer _____ [2]

b) The magnets are rearranged so that the current flowing in the wire is parallel to the field lines.

How will this affect the force on the wire?

_____ [1]

Total Marks _____ / 7

# Induced Potential and Transformers

**1** There are two main types of generators: one produces alternating current and the other produces direct current.

a) Which type of generator produces direct current?    Answer _____ [1]

b) A generator that produces a direct current can be used to power the lights on a bicycle. The rotation of the tyre is used to rotate the coil in the generator.
Using this device as the **only** method of powering a bicycle light is not considered safe.

Suggest why.

_____

_____ [2]

**2** **Figure 1** shows how the output current from an alternator varies with time and the relative position of the magnet within the coil.

Figure 1

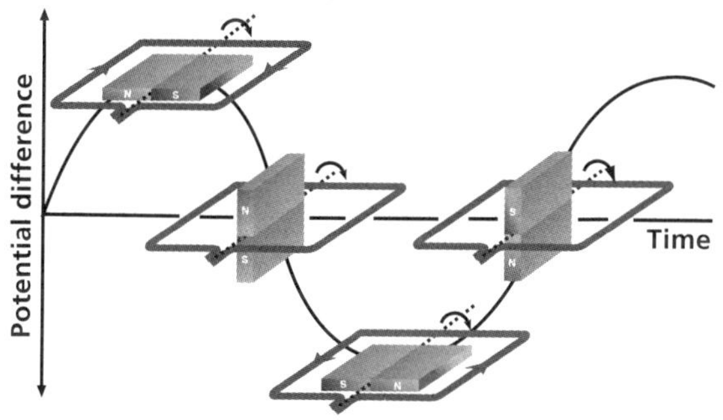

a) In terms of amplitude and frequency, what effect would doubling the speed of rotation have on the graph in **Figure 1**?

_____

_____

_____

_____ [3]

b) In terms of amplitude and frequency, what effect would doubling the strength of the magnet have on the graph in **Figure 1**?

_____

_____ [2]

c) The induced current creates its own magnetic field.

How will this magnetic field affect the force needed to rotate the magnet?

_____

_____ [2]

**Total Marks** _____ / 10

# Particle Model of Matter

**1** Heating a substance can cause it to change state from a solid to a liquid or from a liquid to a gas.

a) What is meant by 'specific latent heat of fusion'?

........................................................................................................................................................

........................................................................................................................................................ [2]

b) While a kettle boils, 0.012kg of water changes to steam.

Calculate the amount of energy required for this change.
Use the correct equation from the Physics Equation Sheet on page 71.
Specific latent heat of vaporisation of water = $2.3 \times 10^6$J/kg

Answer ................................................... [2]

**2** The graph in **Figure 1** shows how temperature varies with time as a substance cools
The graph is **not** drawn to scale.

**Figure 1**

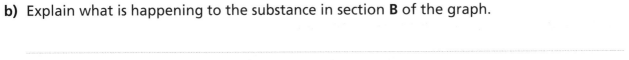

a) Explain what is happening to the substance in section **A** of the graph.

........................................................................................................................................................

........................................................................................................................................................ [2]

b) Explain what is happening to the substance in section **B** of the graph.

........................................................................................................................................................

........................................................................................................................................................

........................................................................................................................................................ [2]

**Total Marks** ................... / 8

# Atoms and Isotopes

**1** Atoms contain three types of particle.

a) Complete the table to show the relative charges of the subatomic particles.

| Particle | Relative Charge |
|---|---|
| Electron | |
| Neutron | |
| Proton | |

[3]

b) A neutral atom has no overall charge.

Explain why in terms of its particles.

........................................................................................................................................

........................................................................................................................................

........................................................................................................................................

[2]

c) Complete the sentences below.

An atom that loses or gains an electron becomes an ............................... .

If it loses an electron, it has an overall ............................... charge.

[2]

**2** In the early part of the 20th century, Rutherford and Marsden investigated the paths taken by positively charged alpha particles into and out of a very thin piece of gold foil. **Figure 1** shows the paths of three alpha particles.

Explain the different paths, **A**, **B** and **C**, of the alpha particles.

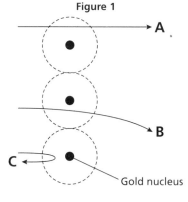

Figure 1

Gold nucleus

........................................................................................................................................

........................................................................................................................................

........................................................................................................................................

........................................................................................................................................

[3]

Total Marks ............... / 10

# Nuclear Radiation

**1** Here is some information about potassium.

> Potassium is a metallic element in Group 1 of the Periodic Table.
> It has an atomic number of 19.
> Its most common isotope is potassium-39, $^{39}_{19}K$.
> Another isotope, potassium-40, $^{40}_{19}K$, is a radioactive isotope.

a) What is meant by 'radioactive isotope'?

_____ [1]

b) During radioactive decay, atoms of potassium-40 change into atoms of a calcium-40.
Calcium-40 has an atomic number of 20 and a mass number of 40.

What type of radioactive decay has taken place?

Answer _____ [1]

c) Potassium-39 does not undergo radioactive decay.

What does this tell us about potassium-39?

_____ [1]

d) Sodium-24 is another radioactive isotope.
It decays by gamma emission.

Give the name of the element formed when this decay takes place.

Answer _____ [1]

**2** Give the unit that is used to measure the activity of a radioactive isotope.

Answer _____ [1]

**3** List the decay mechanisms, **alpha**, **beta** and **gamma**, in order of penetrating power.
Start with the most penetrative.

_____ [1]

Total Marks _____ / 6

# Using Radioactive Sources

**1** Iodine is found naturally in the world and is essential to life.
It is used by the thyroid gland for the production of essential hormones.
Iodine-127 is not radioactive but Iodine-131 is.
Iodine-131 has as a half-life of 8 days.

a) During the Chernobyl nuclear disaster in 1986, an explosion caused a large quantity of the isotope iodine-131 to be released into the atmosphere.

Is iodine-131 from the disaster still a threat to us today?
Explain your answer.

_____

_____

_____ [3]

b) Iodine-131 decays by beta emission and can be used for the treatment of thyroid cancer.

Explain why iodine-131 is suitable for this application.

_____

_____

_____

_____ [4]

c) A sample of iodine-131 has a count-rate of 256 counts per minute.

Work out the count-rate of the sample after 24 days.

Answer _____ [2]

**2** Radioactive isotopes can be used for medical tracers.

Explain how a medical tracer that is ingested in a drink can be used to look for blockages in the intestines.

_____

_____

_____ [3]

Total Marks _____ / 12

# Fission and Fusion

**1** The chart in **Figure 1** shows the sources of background radiation in Britain.

**Figure 1**

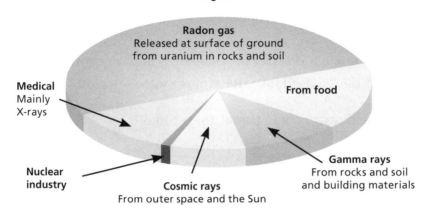

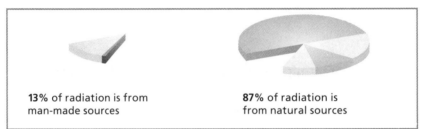

13% of radiation is from man-made sources

87% of radiation is from natural sources

a) Give **two** sources of natural radioactivity from the chart.

........................................................................................................................................ [2]

b) Suggest how the chart might be used to reassure people that nuclear power is safe.

........................................................................................................................................

........................................................................................................................................ [1]

**2** Complete the nuclear equations below by filling in the blanks.

a) Alpha decay: $^{219}Ra \rightarrow\ ^{...}_{84}Po +\ ^{...}_{...}He$ [2]

b) Beta decay: $^{234}Th \rightarrow\ ^{...}_{91}Pa +\ ^{...}_{...}e$ [2]

Total Marks .................. / 7

# Stars and the Solar System

**1** Stars are formed from massive clouds of dust and gases in space.

a) What force pulls the clouds of dust and gas together to form stars?

Answer ................................................................... [1]

b) Once formed, a star can have a stable life for billions of years.

Describe the **two** main forces at work in the star during this period of stability.

.........................................................................................................................................

......................................................................................................................................... [2]

c) What happens to the star once this stable period is over?

.........................................................................................................................................

......................................................................................................................................... [2]

d) Suggest what might happen to a planet close to the star at this time?

......................................................................................................................................... [1]

e) What happens in the final stages of the lifecycle of the largest stars?

......................................................................................................................................... [1]

**2** At the very high temperatures in the Sun, hydrogen is converted into helium.
It takes four hydrogen nuclei to produce one helium nucleus.
**Table 1** shows the relative masses of hydrogen and helium nuclei.

**Table 1**

| Nucleus | Relative Mass |
|---------|---------------|
| Hydrogen | 1.007825 |
| Helium | 4.0037 |

a) Use the data in **Table 1** to calculate by how much the mass of the Sun is reduced every time hydrogen is converted to helium.

Answer ................................................................... [2]

b) Explain how the Sun is able to radiate huge amounts of energy for billions of years but will eventually become a red giant.

.........................................................................................................................................

......................................................................................................................................... [2]

Total Marks .................... / 11

# Orbital Motion and Red-Shift

**1**  A man-made satellite orbits the Earth.
The satellite experiences a resultant force directed towards the centre of the orbit.

    **a)**  What provides the force on the satellite?    Answer _____ [1]

    **b)**  Why does this force not cause the satellite to change speed?

    _____

    _____ [2]

    **c)**  What effect, if any, would changing the speed of a satellite in a stable orbit have?

    _____ [1]

    **d)**  In which direction is the instantaneous velocity of a satellite in orbit?

                                                     Answer _____ [1]

**2**  Red-shift is one of the pieces of evidence that led scientists to propose the 'Big Bang' theory.

    **a)**  Describe the Big Bang theory.

    _____

    _____ [2]

    **b)**  What is meant by red-shift?

    _____

    _____ [2]

    **c)**  Explain how red-shift provides evidence for the Big Bang theory.

    _____

    _____

    _____ [2]

**Total Marks** _____ / 11

# Collins

## GCSE
# PHYSICS
### Paper 1 Higher Tier

**H**

---

**Materials**

Time allowed: 1 hour 45 minutes

> **For this paper you must have:**
> - a ruler
> - a calculator
> - the Physics Equation Sheet (page 71).

### Instructions

- Answer **all** questions in the spaces provided.
- Do all rough work on the page. Cross through any work you do not want to be marked.

### Information

- There are **100** marks available on this paper.
- The marks for each question are shown in brackets [ ].
- You are expected to use a calculator where appropriate.
- You are reminded of the need for good English and clear presentation in your answers.

### Advice

- In all calculations, show clearly how you work out your answer.

01   **Figure 1** shows what happens to each 100 joules of energy from coal that is burned in a power station.

**Figure 1**

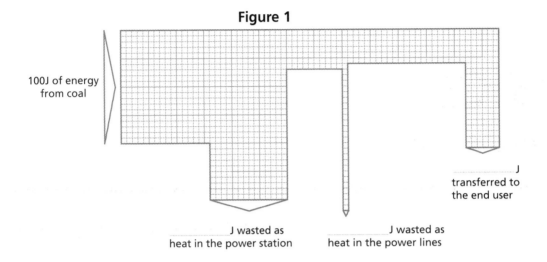

100J of energy from coal

.......................... J
transferred to the end user

.......................... J wasted as heat in the power station

.......................... J wasted as heat in the power lines

**01.1**   Add the missing figures to the diagram.                                              **[3 marks]**

**01.2**   For the same cost, the electricity company could install new power lines that only waste half as much energy as the old ones OR use a quarter of the heat wasted at the power station to heat schools in a nearby town.

Which of these two things do you think they should do?
Give a reason for your answer.

.........................................................................................................................................................

.........................................................................................................................................................

.........................................................................................................................................................

.........................................................................................................................................................

.........................................................................................................................................................                                              **[4 marks]**

**01.3**   Calculate the efficiency of the coal powered station in **Figure 1**.

Efficiency = .............................. %     **[1 mark]**

**02**   A gas burner is used to heat some water in a pan.

By the time the water starts to boil:

- 60% of the energy released has been transferred to the water
- 20% of the energy released has been transferred to the surrounding air
- 13% of the energy released has been transferred to the pan
- 7% of the energy released has been transferred to the gas burner itself.

**02.1**   Some of the energy released by the burning gas is wasted.

What happens to this wasted energy?

................................................................................................................................

................................................................................................................................   **[2 marks]**

**02.2**   What percentage of the energy from the gas is wasted?

Percentage = ........................................ %   **[1 mark]**

**02.3**   How efficient is the gas burner at heating water?

Efficiency = ........................................ %   **[1 mark]**

**Turn over for the next question**

03  A book weighs 6 newtons.
    A librarian picks up the book from the ground and puts it on a shelf that is 2 metres high.

03.1  Calculate the work done on the book.

Work done = _____ J   **[2 marks]**

03.2  The next person to take the book from the shelf accidentally drops it.
      The book falls 2 m to the ground.

      Calculate how much gravitational energy it loses as it falls.
      The gravitational field strength is = 10 N/kg.

Answer = _____ J   **[2 marks]**

03.3  All of the book's gravitational energy is converted to kinetic energy when it falls.

      Calculate the velocity with which the book hits the floor.

Velocity = _____ m/s   **[3 marks]**

**04**  A ripple tank is a piece of lab equipment that can be used to investigate the properties of water waves.

**04.1**  What type of wave is investigated using a ripple tank?

**[1 mark]**

**04.2**  Distinguish between the amplitude and the wavelength of a wave.

**[2 marks]**

**04.3**  Explain how frequency, wavelength and speed of a wave can be measured using a ripple tank. Your answer should consider any cause of inaccuracy in the data.

**[6 marks]**

**04.4**  Explain the importance of controlling the depth of the water in the ripple tank.

**[2 marks]**

**04.5**  Describe a hazard in this investigation and how you would control it.

**[2 marks]**

**Turn over for the next question**

05    A computer is set up to produce a graph of the current through an electric lamp for the first few milliseconds after it is switched on.

The graph is shown in **Figure 2**.

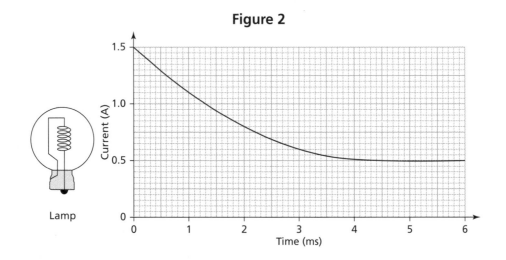

**Figure 2**

05.1    Use the graph to describe how the current through the lamp changes when it is switched on.

........................................................................................................................................

........................................................................................................................................

........................................................................................................................................  **[3 marks]**


05.2    There is a constant 12 V potential difference across the lamp.

What conclusion can be drawn about how the resistance of the lamp changes in the first few milliseconds after it is switched on?

You should state the resistance at different points to support your conclusion.

........................................................................................................................................

........................................................................................................................................  **[3 marks]**

06   A student carries out an experiment to investigate the current through component **X**.
A circuit is set up as shown in **Figure 3**.
The current is measured when different voltages are applied across component **X**.

**Figure 3**

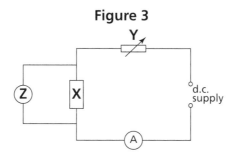

06.1   Name the components labelled **Y** and **Z** in the circuit.

Y = .................................................................................

Z = .................................................................................   [1 mark]

06.2   What is the role of component **Y** in the circuit?

.................................................................................   [1 mark]

**Table 2** shows the measurements obtained in this experiment.

**Table 2**

| Voltage (V)  | −0.6 | −0.4 | −0.2 | 0 | 0.2 | 0.4 | 0.6 | 0.8 |
|--------------|------|------|------|---|-----|-----|-----|-----|
| Current (mA) | 0    | 0    | 0    | 0 | 0   | 50  | 100 | 150 |

06.3   Name the independent variable in this experiment.

Independent variable = .................................................................................   [1 mark]

06.4   Name the dependent variable in this experiment.

Dependent variable = .................................................................................   [1 mark]

**Question 6 continues in the next page**

**06.5** Plot a graph on the axes in **Figure 4** using the data from **Table 1**.

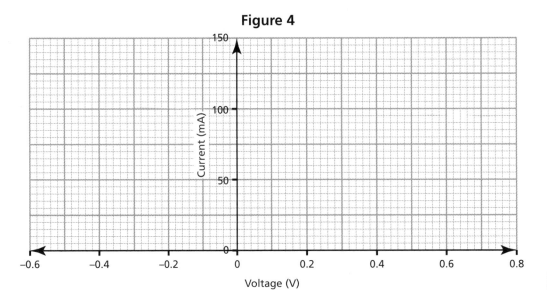

**Figure 4**

[3 marks]

**06.6** The student looks at their measurements and decides that there are no anomalous results.

Are they correct?
You must explain your answer.

[1 mark]

**06.7** Use the shape of the graph to name component **X**.

Component **X** = ................................................ [1 mark]

**07**  A student carries out an experiment to investigate static charge.

Here is the method that they use:

1.  Take a polythene rod, hold it at its centre and rub both ends with a cloth.
2.  Suspend the rod, without touching the ends.
3.  Take a Perspex rod and rub it with another cloth.
4.  Without touching the ends of the Perspex rod, bring each end of the Perspex rod close to each end of the polythene rod.
5.  Make notes on what is observed.

**Figure 5** shows how the apparatus is set up.

**Figure 5**

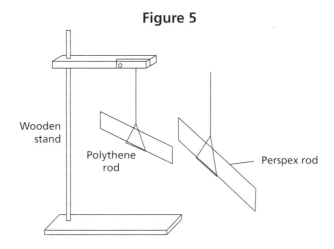

**07.1**  When the Perspex rod is brought close to the polythene rod they attract each other.

Explain why they attract each other.

............................................................................................................................

............................................................................................................  **[2 marks]**

**07.2**  Describe what will happen when the Perspex rod is reversed and the opposite end is brought close to the polythene rod.

............................................................................................................  **[1 mark]**

Question 7 continues in the next page

The experiment is repeated with two polythene rods.

07.3 Describe what will happen when the end of one rod is brought near to the end of the other.

.......................................................................................................................................... [1 mark]

07.4 Explain your answer to 07.3.

..........................................................................................................................................

.......................................................................................................................................... [2 marks]

07.5 Explain, in terms of electron movement, what happens as the rods are rubbed with the cloths.

..........................................................................................................................................

..........................................................................................................................................

..........................................................................................................................................

.......................................................................................................................................... [3 marks]

08    An adaptor can be used to connect up to four appliances in parallel to one 230 V mains socket.

**Table 3** gives a list of appliances and the current that each one draws from a mains socket.

**Table 3**

| Appliance | Current |
|-----------|---------|
| Computer | 1 A |
| Hairdryer | 4 A |
| Heater | 8 A |
| Iron | 6 A |
| Television | 2 A |

08.1  What current will flow to the adaptor when the television, computer and hairdryer are plugged into the adaptor?

Current = _____ A    [1 mark]

08.2  Calculate the electrical power used when the television, computer and hairdryer are plugged into the adaptor.

Power = _____ W    [2 marks]

**Question 8 continues in the next page**

08.3 Explain why it could be dangerous to plug all the devices into the adaptor at the same time.

_____

_____

_____

_____ **[2 marks]**

08.4 Use your answers to **08.1** and **08.2** to calculate the combined resistance when the television, computer and hairdryer are all plugged in and operating.

Resistance = _____ Ω **[2 marks]**

08.5 The television, computer and hairdryer are left running for 5 minutes.

Calculate the charge that flows through the adaptor in this time.

Charge = _____ C **[2 marks]**

09  There are many isotopes of the element strontium (Sr).

09.1  What do the nuclei of different strontium isotopes have in common?

_____

[1 mark]

09.2  The isotope strontium-90 is produced inside nuclear power stations from the fission of uranium-235.

What happens during the process of nuclear fission?

_____

_____

[2 marks]

09.3  When the nucleus of a strontium-90 atom decays, it emits radiation and changes into a nucleus of yttrium-90

$$^{90}_{38}\text{Sr} \rightarrow {}^{90}_{39}\text{Y} + \text{Radiation}$$

What type of decay is this?

Answer _____ [1 mark]

09.4  Give a reason for your answer to **09.3**.

_____

[1 mark]

**Question 9 continues in the next page**

Strontium-90 has a half-life of 30 years

**09.5** What is meant by the term 'half-life'?

......................................................................................................................................... **[1 mark]**

**09.6** After formation in the nuclear reactor, strontium-90 is stored as radioactive waste.

For how many years does strontium-90 have to be stored before its radioactivity has fallen to $\frac{1}{8}$ of its original level?

Answer .................................................... **[2 marks]**

**09.7** In a fission reaction strontium-89 is also produced.
Strontium-89 has a half-life of 50 days.

Explain why this makes it both a more and a less hazardous waste product.

......................................................................................................................................

......................................................................................................................................

...................................................................................................................................... **[2 marks]**

10  A car that is moving has kinetic energy. The faster a car goes, the more kinetic energy it has.

   10.1  The kinetic energy of a car was 472 500 J when travelling at 30 m/s.

      Calculate the total mass of the car.
      Give the unit.

                                                      Mass = ........................................  **[4 marks]**

There is a government road safety campaign to reduce the speed at which people drive in residential areas. It uses the slogan 'Kill your speed, not a child'.
The scientific reason for this is that kinetic energy is transferred from the vehicle to the person it knocks down.

   10.2  A bus and car are travelling at the same speed.
      The bus is likely to cause more harm to a person who is knocked down than the van would.

      Explain why.

      ........................................................................................................................

      ........................................................................................................................

      ........................................................................................................................
                                                                                **[2 marks]**

   10.3  A car and its passengers have a total mass of 1200 kg.
      The car is travelling at 8 m/s.

      Calculate the increase in kinetic energy when the car increases its speed to 14 m/s.

                                      Increase ........................................ J   **[3 marks]**

   10.4  Explain why the increase in kinetic energy is much greater than the increase in speed.

      ........................................................................................................................

      ........................................................................................................................
                                                                                **[1 mark]**

                        **Turn over for the next question**

11    In order to jump over the bar, a high jumper must raise his mass above the ground by 1.25 m. The high jumper has a mass of 65 kg.

The gravitational field strength is 10 N/kg.

**11.1** The high jumper just clears the bar.

Calculate the gain in his gravitational potential energy.

Gain = _____ J    **[2 marks]**

**11.2** Calculate the minimum vertical speed the high jumper must reach in order to jump over the bar.

Use your answer to **11.1** and the formula for kinetic energy.

Minimum vertical speed = _____ m/s    **[3 marks]**

12  The circuit diagram in **Figure 6** shows a circuit used to supply electricity for car headlights.

**Figure 6**

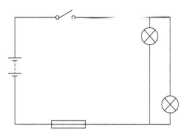

The current through the filament of one car headlight is 2 A.
The potential difference supplied by the battery is 12 V.

**12.1** What is the potential difference across each headlight?

Potential difference = _____ V  **[1 mark]**

**12.2** Work out the total current through the battery.

Current = _____ A  **[1 mark]**

**12.3** Which of the following ratings should be used for the fuse in the circuit?

Tick **one** box.

3 A  ☐

5 A  ☐

10 A  ☐

13 A  ☐                                                                        **[1 mark]**

**12.4** Calculate the resistance of each headlight filament when in use.

Resistance = _____ Ω  **[2 marks]**

**Question 12 continues on the next page**

**12.5** How does the total resistance of the circuit compare to the resistance of each individual bulb?

................................................................................................................................. **[1 mark]**

**12.6** Calculate the power supplied to each of the two headlights of the car.

Power = ................................................ **W** **[2 marks]**

**12.7** The fully charged car battery can deliver 96 kJ of energy at 12 V.

How long can the battery keep both the headlights fully on?

Length of time = ................................................ **s** **[2 marks]**

**END OF QUESTIONS**

# Collins

## GCSE
# PHYSICS
## Paper 2 Higher Tier

**H**

---

**Materials**

Time allowed: 1 hour 45 minutes

**For this paper you must have:**

- a ruler
- a calculator
- a protractor
- the Physics Equation Sheet (page 71).

**Instructions**

- Answer **all** questions in the spaces provided.
- Do all rough work on the page. Cross through any work you do not want to be marked.

**Information**

- There are **100** marks available on this paper.
- The marks for each question are shown in brackets [ ].
- You are expected to use a calculator where appropriate.
- You are reminded of the need for good English and clear presentation in your answers.

**Advice**

- In all calculations, show clearly how you work out your answer.

# Practice Exam Paper 2

**01** A student used a lever system to investigate how the force of attraction between a coil and an iron rocker varied with the current in the coil.
She supported a coil vertically and connected it to an electrical circuit as shown in **Figure 1**.
The weight of the iron rocker is negligible.

**Figure 1**

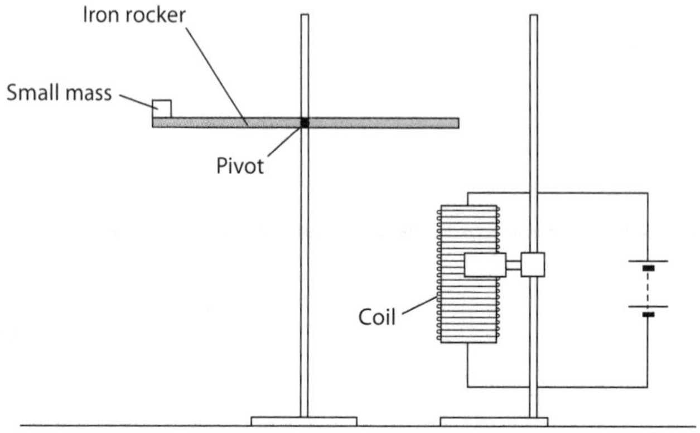


**01.1** Why is it important that the rocker in this experiment is made of iron?

**[1 mark]**

**01.2** The student put a small mass on the end of the rocker and adjusted the current in the coil until the rocker balanced.

To keep the rocker balanced, how will the current through the coil need to change as the size of the mass is increased?

**[1 mark]**

**01.3** Explain your answer to **01.2**.

**[2 marks]**

01.4 A second student set up the same experiment and put an iron core inside the coil.

How will this affect the size of the mass that can be balanced?
You must explain you answer.

[2 marks]

01.5 The student decides to use the original electromagnet and lever system as a force multiplier.
By adjusting the position of the iron bar and the pivot it can be used to lift a larger mass, as shown in **Figure 2**.

**Figure 2**

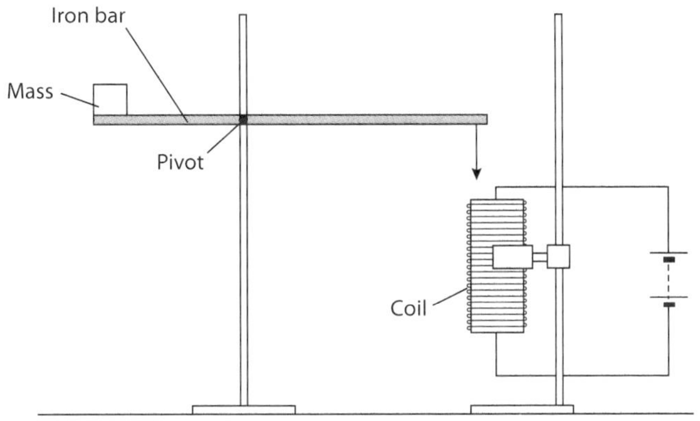

Which of these actions would allow the magnet to lift a heavier mass?

Tick **two** boxes.

Moving the mass closer to the pivot ☐

Moving the magnet closer to the pivot ☐

Reversing the direction of the magnet ☐

Increasing the number of batteries ☐          [2 marks]

**Question 1 continues in the next page**

**01.6** The mass in **Figure 2** has a weight of 50 N.
It is positioned 5 cm from the pivot.

Calculate the turning moment of created by the weight.

Turning moment = _____ **Nm** **[2 marks]**

**01.7** The electromagnet applies a force 20 cm from the pivot,

Use your answer to **01.8** to work out the force needed to exactly balance the weight.

Force = _____ **N** **[2 marks]**

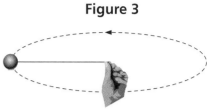

**02**  A group of students investigate circular motion.
They swing a bung attached to a string around in circle.
The string is attached to a force meter, which measures
the centripetal force.

**Figure 3**

The students record how the reading on the force meter
changes to determine how the force affects the speed of the orbit.

**02.1**  In which direction does the centripetal force act on the rubber bung?

............................................................................................................................................ **[1 mark]**

**02.2**  In this experiment, what provides the centripetal force?

............................................................................................................................................ **[1 mark]**

**02.3**  One student swung the rubber bung around in a circle at constant speed.
A second student timed how long it took the rubber bung to complete 10 rotations.

Give **two** variables that are important to control in this experiment.

............................................................................................................................................

............................................................................................................................................ **[2 marks]**

**02.4**  The Moon orbits the Earth in a circular path.

| direction | resistance | speed | velocity |
|---|---|---|---|

Use words from the box to complete the sentences.
You may use each word once, more than once or not at all.
The Moon's ........................... is constant but its ........................... changes.

This is because its ........................... changes. **[1 mark]**

**02.5**  What force provides the centripetal force needed to keep the Moon in its orbit
around the Earth?

............................................................................................................................................ **[1 mark]**

**Turn over for the next question**

03    **Figure 4** shows a transformer.

There is a 50 Hz (a.c.) supply connected to 10 turns of insulated wire wrapped around one side of the iron core.

A voltmeter is connected to 5 turns wrapped around the other side of the iron core.

**Figure 4**

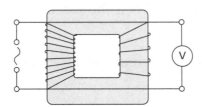

**Table 1** shows values for the potential difference (p.d.) of the supply and the voltmeter reading.

**Table 1**

| Potential Difference of the Supply (V) | Voltmeter Reading (V) |
|---|---|
| 6.4 | 3.2 |
| 3.2 | .............. |
| .............. | 6.4 |

03.1    Complete **Table 1**.    [2 marks]

03.2    Explain in terms of magnetic fields how a transformer works.

.................................................................................................................................

.................................................................................................................................

.................................................................................................................................

.................................................................................................................................

.................................................................................................................................

.................................................................................................................................

.................................................................................................................................

.................................................................................................................................

[4 marks]

**04**   Some students fill an empty plastic bottle with water.
The weight of the water in the bottle is 20 N.
The cross-sectional area of the bottom of the bottle is 0.006 m².

**04.1**   Calculate the pressure of the water on the bottom of the bottle.
Give your answer to 2 significant figures.

Pressure = _____ N/m²    **[3 marks]**

The students made three holes in the bottle along a vertical line: hole A is at the bottom,
B is in the middle, and C is near the top.

**04.2**   From which hole will the water come out at the slowest speed?

Answer = _____    **[1 mark]**

**04.3**   Explain your answer to **4.2**.

_____

_____    **[2 marks]**

**05**   **Figure 5** shows an electromagnetic switch used in a starter motor circuit for a car.

**Figure 5**

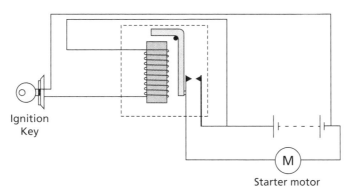

Ignition
Key

M
Starter motor

Explain how turning the ignition key makes a current flow in the starter motor.

_____

_____

_____    **[3 marks]**

**Turn over for the next question**

**06** All stars go through a lifecycle.

About 90% of all stars, including the Sun, are currently in the main sequence period of their lifecycle.

**06.1** Explain briefly how stars like the Sun are thought to have been formed.

........................................................................................................................................

........................................................................................................................................ **[2 marks]**

**06.2** Explain why stars are stable during the main sequence period of their lifecycle.

........................................................................................................................................

........................................................................................................................................ **[1 mark]**

**06.3** The length of time that a star remains in the main sequence of its lifecycle depends on it size and temperature.

Hotter larger stars do not stay in the main sequence for as long as cooler smaller stars.

What does this suggest about the link between the rate of fusion and the size of the star?

........................................................................................................................................

........................................................................................................................................ **[2 marks]**

**06.4** In 1929, the astronomer Edwin Hubble observed that the light from galaxies moving away from the Earth showed a red-shift.

Red-shift provides evidence for the theory that the universe began from one very small initial point.

What name is given to the theory that the universe began in this way?

Answer = ............................................... **[1 mark]**

**06.5** Although the early universe contained only hydrogen, it now contains many different elements.

Describe how the different elements were formed.

........................................................................................................................................

........................................................................................................................................

........................................................................................................................................ **[2 marks]**

07 **Figure 6** shows a coil and a magnet.
An ammeter is connected to the coil.

**Figure 6**

The ammeter has a centre zero scale so that values of current going in either direction through the coil can be measured.

Coil    Magnet

0

Ammeter

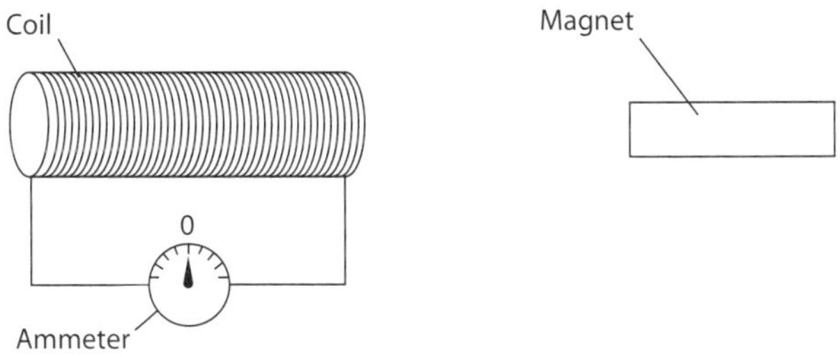

**07.1** When the magnet moves into the coil, the needle on the ammeter moves.

Explain why this happens.

[4 marks]

**Table 2** lists other possible ways of moving the magnet in relation to the coil.

**Table 2**

| Movement of the Magnet | What Happens to the Ammeter Reading? |
|---|---|
| Hold the magnet stationary within the coil. | |
| Move the magnet quickly towards the coil. | |
| Reverse the magnet and move it slowly towards the coil. | |

**07.2** Complete **Table 2** by writing down the effect of each action on the ammeter reading.

[3 marks]

**07.3** The current induced in the solenoid creates a magnetic field.
The north pole of the magnet is moved towards the solenoid.

What pole will be induced on the end of the solenoid closest to the magnet?

Answer = [1 mark]

**Turn over for the next question**

**08** A mobile phone uses a transformer to recharge the battery.
The transformer is connected to a 230 V mains supply and has a 4.6 V output.

**08.1** Explain how you know that this is a step-down transformer?

.................................................................................................................................

................................................................................................................................. **[1 mark]**

**08.2** Describe the construction of a step-down transformer.
You may add to **Figure 7** to help with your answer.

**Figure 7**

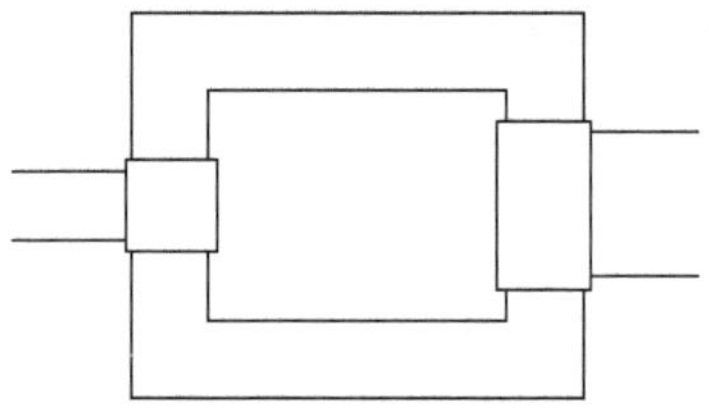

.................................................................................................................................

.................................................................................................................................

.................................................................................................................................

................................................................................................................................. **[4 marks]**

**08.3** The transformer has 2000 turns on the primary coil.

Calculate the number of turns on the secondary coil.
Use the correct equation from the Physics Equation Sheet on page 71.

Answer = ............................................. **[2 marks]**

09    Human ears can detect a range of sound frequencies.

09.1    Complete the sentence.

The range of human hearing is from about ............... Hz to ............... Hz.    **[2 marks]**

09.2    What is ultrasound?

.................................................................................................................................    **[1 mark]**

09.3    The speed of an ultrasound wave in soft tissue in the human body is 1500 m/s.
The frequency of the wave is $2.0 \times 10^6$ Hz.

Calculate the wavelength of the ultrasound wave.

Wavelength = ............................... m    **[2 marks]**

09.4    Describe how ultrasound can be used to find out if a patient is suffering from kidney
stones (a build-up of hard mineral deposits in the kidneys).

.................................................................................................................................

.................................................................................................................................

.................................................................................................................................

.................................................................................................................................    **[2 marks]**

**Turn over for the next question**

**10**   **Figure 8** shows how a convex lens forms an image of an object.
It is **not** drawn to scale.

**Figure 8**

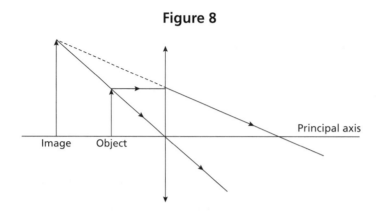

**10.1**   Which of these words can be used to describe the image?
Draw a ring around **two** words.

**diminished      inverted      magnified      real      upright**          **[2 marks]**

**10.2**   The object is 2 cm tall and the image is 6 cm tall.

Work out the magnification of the lens.
Use the correct equation from the Physics Equation Sheet on page 71.

Magnification = ......................................................          **[1 mark]**

11 Three different types of polystyrene were suggested as thermal insulators. Their densities are shown in **Table 3**.

**Table 3**

| Type of Polystyrene | Density (kg/m³) |
|---|---|
| A | 12 |
| B | 18 |
| C | 24 |

**11.1** Predict which material will act as the best thermal insulator. Explain your prediction.

_____

_____

_____

_____ **[2 marks]**

**11.2** Describe an experiment that could be used to test your prediction, including the variables which need to be controlled.

_____

_____

_____

_____

_____

_____

_____

_____ **[6 marks]**

**Question 11 continues in the next page**

**11.3** Give one source of inaccuracy in this investigation and a method of improving it.

[2 marks]

**11.4** Give one hazard of this investigation and a suitable control method.

[2 marks]

12 **Figure 9** shows a lens used as a magnifying glass.

The position of the eye is shown.

The arrow shows the size and position of an object at point **O**.

Use a ruler to accurately construct the position of the image on **Figure 9**.
You should show how you construct your ray diagram and how light appears to come from this image to enter the eye.

**Figure 9**

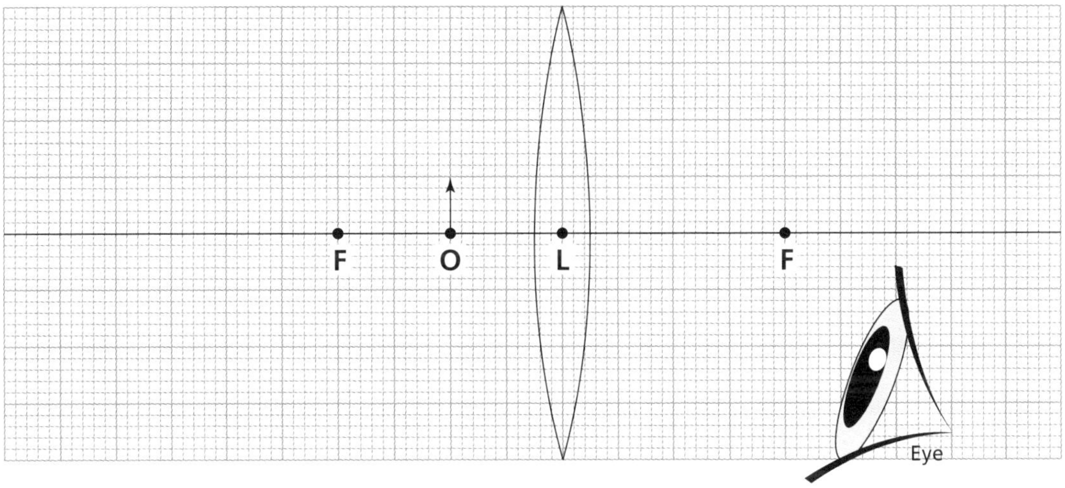

[4 marks]

**Turn over for the next question**

**13** Radio waves, ultraviolet waves, visible light and X-rays are all types of electromagnetic radiation.

**13.1** Choose wavelengths from the list below to complete **Table 4**.

$3 \times 10^{-8}$ m      $1 \times 10^{-11}$ m      $5 \times 10^{-7}$ m      1500 m

**Table 4**

| Type of Radiation | Wavelength (m) |
|---|---|
| Radio waves | |
| Ultraviolet waves | |
| Visible light | |
| X-rays | |

[3 marks]

**13.2** Radio waves can be used to control remote control cars.

Calculate the frequency of radio waves of wavelength 300m.
(The velocity of electromagnetic waves is $3 \times 10^8$ m/s.)

Frequency = ................................................. Hz   [3 marks]

The graph in **Figure 10** shows the speed of a remote-controlled vehicle during a race.

**Figure 10**

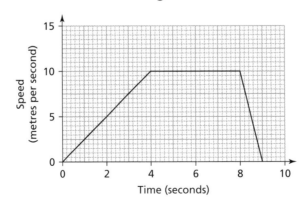

**13.3** Calculate the acceleration during the first four seconds.

Acceleration = _____ m/s² **[3 marks]**

**13.4** What is the maximum speed reached by the vehicle?

Maximum speed = _____ m/s **[1 mark]**

**13.5** How far does the vehicle travel between 4 and 8 seconds?

Distance = _____ m **[2 marks]**

**13.6** At the finish line, a thick wall of rubber foam slows the vehicle down at a rate of 25 m/s².
The vehicle has a mass of 1.5 kg.

Calculate the average force of the rubber foam on the car.

Average force = _____ N **[2 marks]**

**Turn over for the next question**

**14**   When a gun is fired, a very large force acts on the bullet for a very short time.
An average force of 4000 newtons acts for 0.004 seconds on a bullet of mass 50 g.

   **14.1**   Calculate the momentum gained by the bullet.
   Use the correct equation from the Physics Equation Sheet on page 71.

Momentum = .................................................... kg m/s   **[2 marks]**

   **14.2**   Calculate the speed of the bullet.

Speed = .................................................... m/s   **[2 marks]**

**END OF QUESTIONS**

# Physics Equation Sheet

| | | |
|---|---|---|
| 1 | pressure due to a column of liquid = height of column × density of liquid × gravitational field strength (*g*) | $p = h\rho g$ |
| 2 | (final velocity)² − (initial velocity)² = 2 × acceleration × distance | $v^2 - u^2 = 2as$ |
| 3 | force = $\dfrac{\text{change in momentum}}{\text{time taken}}$ | $F = \dfrac{m\Delta v}{\Delta t}$ |
| 4 | elastic potential energy = 0.5 × spring constant × (extension)² | $E_e = \frac{1}{2}ke^2$ |
| 5 | change in thermal energy = mass × specific heat capacity × temperature change | $\Delta E = mc\Delta\theta$ |
| 6 | period = $\dfrac{1}{\text{frequency}}$ | |
| 7 | magnification = $\dfrac{\text{image height}}{\text{object height}}$ | |
| 8 | force on a conductor (at right-angles to a magnetic field) carrying a current = magnetic flux density × current × length | $F = BIl$ |
| 9 | thermal energy for a change of state = mass × specific latent heat | $E = mL$ |
| 10 | $\dfrac{\text{potential difference across primary coil}}{\text{potential difference across secondary coil}} = \dfrac{\text{number of turns in primary coil}}{\text{number of turns in secondary coil}}$ | $\dfrac{V_p}{V_s} = \dfrac{n_p}{n_s}$ |
| 11 | potential difference across primary coil × current in primary coil = potential difference across secondary coil × current in secondary coil | $V_s I_s = V_p I_p$ |
| 12 | For gases: pressure × volume = constant | $pV = constant$ |

# Answers

## Topic-Based Questions

### Page 4 Forces – An Introduction

1. a) Weight / gravity [1]; non-contact [1]; air resistance (or drag) [1]; contact [1]
   b) weight = mass × gravitational field strength / $W = mg$ [1]; weight = 120 000 × 10 = 1 200 000N [1]
   c) A correctly drawn horizontal arrow [1]; a vertical arrow that is half the length of the horizontal arrow [1]; and a diagonal arrow showing the total resultant force [1]

Resultant Force

2. a) Friction [1]
   b) Friction is a contact force [1]; lifting the glider off the track means there is no contact, so no friction [1]

### Page 5 Forces in Action

1. a) Remove the weights [1]; if it returns to its original shape it was behaving elastically [1]
   b) i) 8.0 (cm) [1]
      ii) The extension appears to be linear [1]; and is increasing by 4cm each time [1]
2. a) 200 × 1.6 [1]; = 320N [1]
   b) Tree moment = man moment [1]; 320 = 0.2 × F [1]; $F = \frac{320}{0.2}$ = 1600N [1]

### Page 6 Pressure and Pressure Differences

1. a) pressure = $\frac{\text{force normal to surface}}{\text{area of that surface}}$ / $p = \frac{F}{A}$ [1]
   b) The man presses down on a small area [1]; this increases the pressure of the liquid [1]; at the car the pressure acts on a bigger area and $F = pA$ so the force is increased [1]
2. a) The balloon is less dense than the surrounding air [1]
   b) At a higher altitude the air is less dense [1]; this means the force of upthrust on the balloon falls as it rises [1]; eventually upthrust and weight are balanced, so there is no resultant vertical force on the balloon [1]

### Page 7 Forces and Motion

1. a) Zero [1]

   Displacement is the distance from the start point. The person returns home – it is a circular journey – so the total displacement at the end of their journey is zero.

b) B [1]
c) Stationary [1]
d) A is travelling away from home [1]; D is travelling in the opposite direction (back towards home) [1]; D is travelling slightly faster than A [1]

2. a) speed = $\frac{\text{distance travelled}}{\text{time}}$
   $v = \frac{s}{t}$ [1]; = $\frac{180}{6}$ = 30m/s [1]
   b) 50 – 40 = 10m/s [1]
   c) 50 + 40 = 90m/s [1]

   The velocity changes from 50m/s to the east to 40m/s to the west.

### Page 8 Forces and Acceleration

1. a) The mass of the system [1]
   b) The force accelerating the trolley (provided by the hanging masses) [1]
   c) The acceleration of the trolley [1]

   The independent variable is the one deliberately changed and the dependent variable is the one being measured.

2. a) acceleration = $\frac{\text{change in velocity}}{\text{time taken}}$
   $a = \frac{\Delta v}{t}$ [1]; acceleration = $\frac{16 - 4}{8}$ = 1.5 [1]; m/s$^2$ [1]
   b) resultant force = mass × acceleration / $F = ma$ [1]; force = 68 × 1.5 = 102N [1]
3. a) acceleration = $\frac{33}{11}$ [1]; = 3m/s$^2$ [1]
   b) force = 950 × 3 [1]; = 2850N [1]

### Page 9 Terminal Velocity and Momentum

1. They could change position e.g. dive head first [1]; to become more aerodynamic / reduce air resistance [1]
2. a) force = $\frac{\text{change in momentum}}{\text{time taken}}$
   $F = \frac{m\Delta v}{\Delta t}$, 6 700 000 = (5000 × v) – (5000 × 0) [1];
   $v = \frac{6\,700\,000}{5000}$ [1]; = 1340m/s [1]

   change in momentum = final momentum – initial momentum = $mv - mu$

b) mass burned = 5000 × 5 × 10 = 250 000kg [1]; momentum given to fuel = 250 000 × 1340 = 335 000 000kg m/s [1]; spacecraft gains 335 000 000kg m/s [1]

   Momentum is conserved, so the momentum gained by the rocket is equal to the momentum given to the fuel.

c) momentum = mass × velocity / $p = mv$, 335 000 000 = 3 000 000 × v [1];
   $v = \frac{335\,000\,000}{3\,000\,000}$ = 111.67m/s [1]
d) As the rocket burns fuel, it becomes lighter [1]; resultant force = mass × acceleration / $F = ma$, so acceleration ($a$) increases [1]; at a higher altitude, there is less air resistance [1]; so the resultant force increases [1]

### Page 10 Stopping and Braking

1. a) 1.4 seconds [1]
   b) 1.4 × 15 [1]; = 21m [1]
   c) 4 – 1.4 = 2.6 seconds [1]
   d) momentum = mass × velocity / $p = mv$ [1]; 1300 × 15 = 19 500kg m/s [1]
   e) force = rate of change of momentum [1]; force = $\frac{19\,500}{2.6}$ [1]; = 7500N [1]
   f) A correctly drawn graph line that starts horizontally at 15m/s [1]; then starts sloping downwards between 0.2s and 0.8s [1]; and has the same gradient on the downslope as the original line [1]

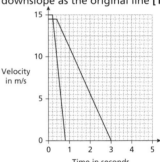

   g) **Any two of:** The down slope would start at the same point [1]; but have a shallower gradient [1]; and take a longer total time to stop [1]

### Page 11 Energy Stores and Transfers

1. a) 100 – 20 = 80 degree change [1]; energy = 2 × 4200 × 80 [1]; = 672 000J [1]
   b) 672 000 = 2.5 × 4200 × temp change [1]; temp change = $\frac{672\,000}{(2.5 \times 4200)}$ = 64°C [1]; final temp = 20 + 64 = 84°C [1];
2. a) gravitational potential energy = mass × gravitational field strength × height / $E_p = mgh$ [1]; $E_p$ = 0.1 × 10 × 0.05 [1]; = 0.05J [1]
   b) kinetic energy = 0.5 × mass × (speed)$^2$ / $E_k = \frac{1}{2}mv^2$ [1]; 0.05 = 0.5 × 0.1 × $v^2$ [1]; $v^2 = \frac{0.05}{(0.5 \times 0.1)}$ = 1, $v$ = 1m/s [1]

## Page 12 Energy Transfers and Resources

1. light [1]; electrical / thermal [1]; heat [1]
2. a) Start temperature of water [1]; thickness of fleece [1]
   b) Fleece M [1]; because it cools the slowest, so insulates the best [1]

## Page 13 Waves and Wave Properties

1. a) Half a wave per second [1] (Accept: 1 wave every 2 seconds)
   b) wave speed = frequency × wavelength / $v = f\lambda$ [1]
   c) speed = $\dfrac{\text{distance}}{\text{time}}$ / $v = \dfrac{s}{t}$, $v = \dfrac{50}{10}$ [1]; = 5m/s [1]
   d) $v = f\lambda$, $\lambda = \dfrac{5}{0.5}$ [1]; = 10m [1]
   e) They will go slower [1]
2. In longitudinal waves, the particles oscillate [1]; parallel to the direction of energy transfer / wave motion [1]; in transverse waves, the oscillation is at right-angles to the direction of energy transfer / wave motion [1]

## Page 14 Reflection, Refraction and Sound

1. Can be shown on the diagram to help explain but must include 'refraction' in the answer for full marks, e.g. Light rays from the pin [1]; are refracted when they leave the water [1]; away from the normal and into the eye [1]

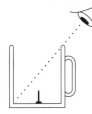 

2. a) Two lines with arrows drawn to show reflection[1]; with the angle of incidence equal to the angle of reflection [1]

   b) Will make it quieter [1]; because the carpet absorbs the sound energy [1]

## Page 15 Waves for Detection and Exploration

1. distance travelled = speed × time / $s = vt$ [1]; $s = 1600 \times 0.8 = 1280$m [1]; $\dfrac{1280}{2} = 640$m [1]

2. a) The back of the steel block [1]
   b) 90mm [1]
   c) It is smaller / half the size [1]

## Page 16 The Electromagnetic Spectrum

1. a) Microwaves [1]
   b) Accept any sensible answer, e.g. X-rays for photographing bones OR gamma rays for sterilisation OR UV for sunbeds [2] (1 mark for wave, 1 mark for use)
2. frequency [1]; wavelength [1]
3. a) X-rays [1]
   b) It can penetrate soft tissue [1]; but is blocked by bone [1]

## Page 17 Lenses

1. a) magnification = $\dfrac{2}{8}$ [1]; = 0.25 [1]
   b) magnification = $\dfrac{300}{2}$ [1]; = 150 [1]

   Make sure both measurements are in the same units before calculating magnification.

2. Parallel rays of light entering the lens [1]; spread out / diverge when they leave the lens [1]
3. a) A convex / converging lens [1]
   b) More distant from the lens than the object [1]
4. a) Convex [1]
   b) Three correctly drawn ray lines with arrows [3] (1 mark for each line); correctly drawn image [1]

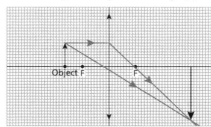

   c) magnification = $\dfrac{\text{height of image}}{\text{height of object}}$ = $\dfrac{20}{10}$ [1]; = 2 [1] (Allow correct calculation with readings from own diagram if drawn incorrectly)

## Page 18 Light and Black Body Radiation

1. a) The red filter only lets red light through [1]; so only red light reaches the green filter, which only lets green light through, so no light gets through [1]
   b) A green circle [1]
2. a) That only infrared light is emitted / its wavelength it too long to be in the visible spectrum [1]
   b) It is cooler than the Sun [1]
   c) Radiation emitted increases in frequency as temperature increases [1]; only low frequency infrared is emitted [1]; so must be a lower temperature than the Sun, which emits higher frequency visible light [1]

## Page 19 An Introduction to Electricity

1. current [1]; charge [1]; greater [1]; current [1]
2. a) potential difference = current × resistance / $V = IR$ [1]
   b) $230 = 5 \times R$ [1]; $R = \dfrac{230}{5} = 46\Omega$ [1]
   c) charge = current × time / $Q = It$ [1]
   d) $Q = 5 \times 120$ [1]; = 600C [1]
3. Energy [1]; greater [1]; current [1]
4. a) Closed switch [1]
   b) Battery [1]
   c) Fuse [1]
   d) Light dependent resistor / LDR [1]

## Page 20 Circuits and Resistance

1. a) To adjust the resistance of the circuit [1]; and change the voltage across the component [1]
   b) Series [1]
   c) Parallel [1]
2. Four correctly drawn lines [3] (2 marks for two correct lines; 1 mark for one correct line)
   Light dependent resistor (LDR) – Resistance decreases as light intensity increases.
   Thermistor – Resistance decreases as temperature increases.
   Diode – Has a very high resistance in one direction.
   Filament light – Resistance increases as temperature increases.

## Page 21 Circuits and Power

1. a) energy transferred = power × time / $E = Pt$ [1]; $E = 1600 \times 120$ [1]; = 192 000J [1]
   b) $\dfrac{192\,000}{100} \times 10$ [1]; = 19 200J [1] (accept any equivalent method)
2. a) $V = IR$, $V = 2 \times 3$ [1]; = 6V [1]
   b) $18 = 6 + V$ [1]; $V = 18 - 6 = 12$V [1]

   The total potential difference in a series circuit is shared across the components.

   c) $18 = 2 \times R$ [1]; $R = \dfrac{18}{2} = 9\Omega$ [1]

## Page 22 Domestic Uses of Electricity

1. a) 1.5V [1]; d.c. [1]
   b) 230V [1]; a.c. [1]
2. When a device is switched off, the live wire before the switch can still be at a non-zero potential [1]; touching this would create a potential difference between the wire and the ground [1];

# Answers

this would make current flow through the person [1]; which would cause an electric shock [1]

## Page 23 Electrical Energy in Devices

1. a) Kinetic [1]
   b) It is dissipated / lost [1]; to the surroundings [1]
2. a) Energy input = electrical [1]; useful energy output = kinetic [1]
   b) It disappears. [1]
   c) 500J [1]
3. The output from the generators goes through a step-up transformer [1]; this increases the voltage and also reduces the current [1]; the low current stops the cables from becoming hot [1]; which means less energy is lost during transmission [1]

## Page 24 Static Electricity

1. a) True [1]
   b) False [1]
   c) False [1]
   d) True [1]
2. a) Friction [1]; has caused electrons to be removed from the rod / transferred to the jumper [1]; leaving the rod positively charged [1]
   b) repel [1]; attract [1]
3. charged [1]; electric [1]; force [1]; force [1]; charges [1]

## Page 25 Magnetism and Electromagnetism

1. If free to move, the magnet will rotate so that the north pole of the magnet [1]; points to the Earth's north pole [1]
2. a) They will repel [1]
   b) It will be attracted to the magnet [1]
3. When the switch is pressed current flows in the electromagnet [1]; this magnetises the magnet [1]; which attracts the armature, causing the hammer to hit the gong [1]; the movement of the armature breaks the circuit, switching off the magnet [1]; the armature springs back and remakes the circuit, which starts the cycle again [1]

## Page 26 The Motor Effect

1. The alternating electrical signal passes through the coil between the poles of the magnet [1]; which creates an alternating force on the coil [1]; the coil is fixed to the cone, so there is an alternating forces on the cone [1]; which vibrates at the same frequency as the alternating signal [1]

2. a) force on a conductor (at right-angles to a magnetic field carrying a current = magnetic flux density × current × length / $F = BIl$, $F = (0.3 \times 10^{-3}) \times 2 \times 0.05$ [1]; = 0.00 003N [1]
   b) There will be no force on the wire [1]

## Page 27 Induced Potential and Transformers

1. a) A dynamo [1]
   b) The electricity produced is not constant [1] so the light brightness varies/gets dim at low speed/ goes off when stopped [1]
2. a) The amplitude of the waves would be twice as high [1]; and the frequency of the waves would be doubled [1]; so the period would be halved [1]
   b) The amplitude of the waves would be twice as high [1]; and the frequency of waves would be unchanged [1]
   c) The induced current will oppose the motion [1]; so the force needed to keep the alternator turning increases [1]

## Page 28 Particle Model of Matter

1. a) The energy required [1]; to change 1kg of a substance from a solid to a liquid [1]
   b) thermal energy for a change of state = mass × specific latent heat / $E = mL$, $E = 0.012 \times (2.3 \times 10^6)$ [1]; = 27 600J [1]
2. a) The substance is condensing [1]; from a gas to a liquid [1]
   b) The particles are slowing down [1]; and the substance is cooling [1]

## Page 29 Atoms and Isotopes

1. a) Electron, –1 [1]; neutron, 0 [1]; proton, +1 [1]
   b) It has the same number of protons [1]; as electrons [1]
   c) ion [1]; positive [1]
2. Path A is a long way from the nucleus and the alpha particle goes straight through [1]; Path B is close to the positive nucleus so the alpha particle is deflected [1]; Path C comes very close to the nucleus and the alpha particle is repelled back the way it came [1]

Two positively charged particles will repel each other.

## Page 30 Nuclear Radiation

1. a) An unstable atom that gives out radiation [1]
   b) Beta decay [1]

   c) It is stable / non-radioactive [1]
   d) Sodium [1]
2. Becquerel [1]
3. Gamma, beta, alpha [1]

## Page 31 Using Radioactive Sources

1. a) No longer a risk [1]; because it has a half-life of just 8 days [1]; so would have completely decayed to the same level as background radiation in the last 30 years [1]
   b) Iodine-131 has a short half-life [1]; so it doesn't remain radioactive in the body for long [1]; Iodine-131 gathers in the thyroid so stays close to the cancer [1]; and beta radiation then destroys the cancer cells, as it is moderately ionising [1]
   c) 24 days is 3 half-lives [1]; 256 → 128 → 64 → 32, so a count rate of 32 remains [1]
2. It travels into the intestines and cannot pass the blockage [1]; a detector outside the body can detect the radiation [1]; no radiation will be detected after the blockage, so the location of the blockage can be found [1]

## Page 32 Fission and Fusion

1. a) **Any two of:** rocks/soil [1]; radon gas [1]; food [1]; cosmic rays [1]
   b) The nuclear industry contributes a very small percentage of the total background radiation – much less than natural sources [1] (Accept any other sensible answer)
2. a) Alpha decay:
   $$^{219}_{86}\text{Ra} \rightarrow\, ^{215}_{84}\text{Po} + ^{4}_{2}\text{He}\,\text{[2]}$$
   (1 mark for the correct radioactive particle on the left-hand side of the equation; 1 mark for the correct products)
   b) Beta decay:
   $$^{234}_{90}\text{Th} \rightarrow\, ^{234}_{91}\text{Pa} + ^{0}_{-1}\text{e}\,\text{[2]}$$
   (1 mark for the correct radioactive particle on the left-hand side of the equation; 1 mark for the correct products)

## Page 33 Stars and the Solar System

1. a) Gravity [1]
   b) The explosive force of fusion / radiation force [1]; and the compressive force of gravity [1]
   c) It expands [1]; to become a red giant [1]
   d) It could be engulfed by the star [1]
   e) They explode as a supernova and become a black hole or neutron star [1]
2. a) (1.007 825 × 4) – 4.0037 [1]; = 0.0276, so the relative mass falls by 0.0276 [1]

b) Eventually the decrease in mass will be significant enough [1]; that the gravitational attraction due to the Sun's mass will decrease, allowing the material to expand (as a red giant) [1]

## Page 34 Orbital Motion and Red-Shift

1. a) The gravitational attraction of the Earth [1]
   b) Because the satellite is in circular motion [1]; the acceleration caused by the force makes the direction change, but the magnitude (speed) is unchanged [1]
   c) The radius of the orbit would change [1]
   d) At right-angles to the force of gravity [1]
2. a) The universe started with an explosion from a single point [1]; and has been expanding ever since [1]
   b) The light from a light source moving away becomes redder [1]; as its wavelength is stretched and becomes longer [1]
   c) All of the distant galaxies are moving away from one another [1]; and they appear to be moving away from the same place [1]

## Pages 35–52 Practice Exam Paper 1

01.1 From left to right: 65J [1]; 5J [1]; 30J [1]

01.2 Heat the schools [1]; because it saves the most energy [1]; half of 5J is 2.5J [1]; but a quarter of 65J is over 15J [1]

01.3 efficiency =

$$\frac{\text{useful output energy transfer}}{\text{total input energy transfer}} \times 100\%,$$

$$= \frac{30}{100} \times 100\% = 30\% \quad [1]$$

02.1 It spreads out / is dissipated [1]; into the surroundings [1]

02.2 20 + 13 + 7 = 40% [1]

02.3 60% [1]

03.1 work done = force × distance / $W = Fs$, $W = 6 \times 2$ [1]; = 12J [1]

03.2 weight = mass × gravitational field strength / $W = mg$, $m = \frac{6}{10} = 0.6$kg [1]; gravitational potential energy = mass × gravitational field strength × height / $E_p = mgh$, $E_p = 0.6 \times 10 \times 2 = 12$J [1]

03.3 kinetic energy = 0.5 × mass × (speed)² / $E_k = \frac{1}{2}mv^2$ [1]; $v^2 = \frac{12}{0.5 \times 0.6} = 40$ [1]; $v = \sqrt{40} = 6.3$m/s [1]

04.1 A transverse wave [1]

04.2 The amplitude of a wave is the maximum displacement of a point on a wave away from its undisturbed position [1]; the wavelength of a

wave is the distance from a point on one wave to the equivalent point on the adjacent wave [1].

04.3 **Here is a sample answer worth 6 marks:** Produce a wave in the ripple tank. Time how long it takes this wave to travel the length of the tank. Use this time to calculate the wave speed using the formula speed = distance ÷ time. To find the frequency, count the number of waves which pass a fixed point in a given time (e.g. 10 seconds). Divide this number by the time to give the frequency. Use a ruler to estimate the distance between the peaks of the wave as it travels. Repeat the experiment several times and then use the repeat readings to calculate mean values for the frequency, wavelength and speed. Sources of inaccuracy include it being difficult to count and measure small or fast waves. Estimating the distance with a ruler is also an inaccurate way to determine wavelength. A stroboscope could be used in the investigation to improve accuracy.

04.4 The water depth will affect the wave speed and wavelength at a given frequency [1]; so changes in depth along the tank will lead to changes in the speed and wavelength as the wave moves along the tank [1].

04.5 Water could splash out of the tank causing a potential slip hazard [1]; control this by ensuring that, when waves are produced, water does not splash out and any spills are wiped up immediately [1]

Or Using a stroboscope presents a risk to people with photosensitive epilepsy [1]; to control this, ensure that no one at risk is in the room when the stroboscope is used [1].

Or **Any other valid hazard and control.**

05.1 The current starts at 1.5A [1]; decreases for 4ms [1]; then stabilises at 0.5A [1]

05.2 The resistance starts low at 8 ohms [1]; it increases rapidly [1]; until it reaches 24 ohms [1]

06.1 Y = variable resistor; Z = voltmeter [1]

06.2 To control the voltage applied to component X / adjust the resistance [1]

06.3 The voltage [1]

06.4 The current [1]

06.5 Accurately plotted voltage [1]; and current [1]; with all points joined by a smooth curve [1]

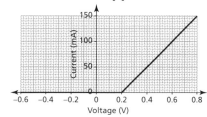

06.6 Yes, they are correct as all the points fit the line [1]

An anomalous result would be significantly higher or lower than the other results or would not fit the pattern.

06.7 A diode [1]

07.1 When rubbed they became charged [1]; they are made of different materials and have opposite charges, which attract [1]

07.2 They still attract [1]

07.3 They will repel [1]

07.4 They are made of the same material, so end up with the same charge [1]; and like charges repel [1]

07.5 When an insulating material is rubbed, electrons move [1]; either from the rod to the cloth, leaving the rod positive [1]; or from the cloth to the rod, making the rod negative [1]

08.1 2 + 1 + 4 = 7A [1]

08.2 $P = 230 \times 7$ [1]; = 1610 [1]

08.3 This could damage / overload the circuit [1]; and cause an electric shock [1]

08.4 power = (current)² × resistance / $P = I^2R$, $1610 = 7^2 \times R$ [1]; $R = \frac{1610}{49} = 32.9\Omega$ [1]

08.5 charge flow = current × time / $Q = It$, $Q = 7 \times 60 \times 5$ [1]; = 2100C [1]

09.1 They have the same number of protons [1]

09.2 A large unstable nucleus [1]; splits into two or more smaller nuclei and releases energy [1]

09.3 Beta decay [1]

09.4 The mass number has not changed during emission, but the proton number has increased by one [1]

09.5 The time it takes for half of the radioactive isotopes to decay / for the count rate to halve [1]

09.6 $\frac{1}{8}$ means that 3 half-lives have passed [1]; so is (30 × 3 =) 90 years [1]

09.7 It is more hazardous because it decays quickly / is highly active, so gives out radiation quickly [1]; but is not radioactive for a long time, so in the long-term it is less hazardous [1]

10.1 kinetic energy = 0.5 × mass × (speed)² / $E_k = \frac{1}{2}mv^2$, $m = \frac{E_k}{\frac{1}{2}v^2}$ [1]; $m = \frac{472500}{450}$ [1]; = 1050 [1]; final answer = 1050kg (unit must be correctly stated for mark) [1]

10.2 The bus has more mass [1]; so it has more kinetic energy [1]

10.3 kinetic energy = 0.5 × mass × (speed)² / $E_k = \frac{1}{2}mv^2$ [1]; increase in $E_k = 0.5 \times 1200 \times (14^2 - 8^2)$ [1]; = 79 200J [1]

10.4 Because the kinetic energy depends on the square of the speed [1]

# Answers

**11.1** gravitational potential energy = mass × gravitational field strength × height / $E_p = mgh$,
$E_p = 65 \times 10 \times 1.25$ **[1]**; = 812.5J **[1]**

**11.2** kinetic energy =
0.5 × mass × (speed)$^2$ / $E_k = \frac{1}{2}mv^2$,
$812.5 = 0.5 \times 65 \times v^2$ **[1]**; $v^2 = 25$ **[1]**;
$v = \sqrt{25}$ = 5m/s **[1]**

**12.1** 12V **[1]**
**12.2** 2 + 2 = 4A **[1]**
**12.3** 5A **[1]**
**12.4** potential difference = current × resistance / $V = IR$,
$R = \frac{12}{2}$ **[1]**; = 6Ω **[1]**
**12.5** It is less / half **[1]**
**12.6** power = potential difference × current / $P = VI$, $P = 12 \times 2$ **[1]**; = 24W each **[1]**
**12.7** $\frac{96\ 000}{(24 \times 2)}$ **[1]**; = 2000s **[1]**

## Pages 53–70 Practice Exam Paper 2

**01.1** So it will be attracted by the magnetic coil **[1]**
**01.2** It will need to increase **[1]**
**01.3** A bigger mass makes a bigger force pulling down on the left, so a bigger force is needed on the right **[1]**; a bigger current will increase the strength of the magnetic field created by the coil **[1]**
**01.4** It will increase the mass that can be balanced **[1]**; because the iron core will make the magnet stronger **[1]**
**01.5** Moving the mass closer to the pivot **[1]**; Increasing the number of batteries **[1]**
**01.6** Moment of force = force × distance / $M = Fd$, $M = 50 \times 0.05$ **[1]**; = 2.5Nm **[1]**
**01.7** $2.5 = F \times 2$ **[1]**; $F = \frac{2.5}{0.2}$ = 12.5N **[1]**
**02.1** Towards the centre of the circle **[1]**
**02.2** The tension of the string **[1]**
**02.3** The radius of the circle **[1]**; and the mass of the bung **[1]**
**02.4** speed; velocity; direction **[1]**
**02.5** Gravity / the gravitational attraction of the Earth on the Moon **[1]**
**03.1** Missing voltmeter reading = 1.6V **[1]**; missing potential difference of supply = 12.8V **[1]**
**03.2** An alternating potential difference **[1]**; across the primary coil **[1]**; induces an alternating magnetic field in the core **[1]**; this alternating field induces an alternating potential difference in the secondary coil **[1]**
**04.1** pressure =
$\frac{\text{force normal to the surface}}{\text{area of that surface}}$ /
$P = \frac{F}{A}$ **[1]**; $P = \frac{20}{0.006}$ = 3333.33 **[1]**;
= 3300N/m$^2$ **[1]**
**04.2** Hole C **[1]**

**04.3** It is has the least weight of water above it **[1]**; so is the lowest pressure **[1]**
**05** Turning the key switches the electromagnet on **[1]**; this attracts the pivoted armature **[1]**; which pushes the contacts together, completing the starter motor circuit **[1]**
**06.1** Gravity pulls clouds of hydrogen gas together **[1]**; once there is enough mass the star becomes hot and dense enough for fusion to start **[1]**
**06.2** The explosive / expansive forces of fusion are balanced **[1]**; by the compressive / attractive force of gravity **[1]**
**06.3** The bigger the star the faster the rate of fusion **[1]**
**06.4** The Big Bang theory **[1]**
**06.5** In stars, larger elements are formed during fusion **[1]**; the largest stars can fuse bigger elements and the heaviest elements are formed during a supernova **[1]**
**07.1** As the magnet moves into the coil **[1]**; the magnetic field lines are cut by the coil **[1]**; this induces a potential difference in the coil **[1]**; and because the coil is part of a complete circuit a current flows, which is measured on the ammeter **[1]**
**07.2** **From top to bottom:** ammeter reads zero **[1]**; ammeter reading is high **[1]**; ammeter reading is low in the opposite direction to the original situation **[1]**
**07.3** A north pole **[1]**
**08.1** The output voltage is lower than the input voltage **[1]**
**08.2** The transformer is made from an iron core **[1]**; the input voltage is supplied to the primary coil **[1]**; and the output voltage induced in the secondary coil **[1]**; there are more turns on the primary coil than on the secondary coil **[1]**
**08.3** $\frac{\text{potential difference across primary coil}}{\text{potential difference across secondary coil}}$
$= \frac{\text{number of turns in primary coil}}{\text{number of turns in secondary coil}}$'
$\frac{230}{4.6} = \frac{2000}{\text{number of turns in secondary coil}}$
**[1]**; number of turns in secondary coil
$= \frac{2000}{50}$ = 40 **[1]**
**09.1** 20Hz **[1]**; 20000Hz **[1]**
**09.2** Sound with a frequency greater than 20 000Hz **[1]**
**09.3** wave speed = frequency × wavelength / $v = f\lambda$,
$\lambda = \frac{1500}{2\ 000\ 000}$ **[1]**; = 0.00075m **[1]**
**09.4** The ultrasound is transmitted into the body and is reflected back if kidney stones are present **[1]**; the time delay before the reflected

wave is detected by the receiver indicates where the stones are **[1]**
**10.1** magnified **[1]**; upright **[1]**
**10.2** 3 **[1]**
**11.1** Material A as it has the lowest density **[1]**; therefore it should be the best insulator as it contains more air spaces / will conduct the heat slowest **[1]**.
**11.2** **Here is a sample answer worth 6 marks:** Wrap a test tube or beaker in material A. Fill the test tube or beaker with hot water at a certain temperature (e.g. 60°C). Start a stop watch and record the temperature of the water every minute until the temperature stops decreasing (reaches room temperature). Record the time taken for the water to reach room temperature. Repeat the investigation two more times and use the repeat results to calculate a mean time taken. Repeat the whole investigation for materials B and C, including taking repeat readings. Two variables that need to be controlled in this investigation are the thickness of the material and the volume of water used.
**11.3** The insulation material may not totally cover the beaker or test tube, so heat may be lost through holes or gaps **[1]**; control this by ensuring the insulating material completely covers the beaker or test tube with only the thermometer poking out **[1]**.
**Or** Time interval used may not be enough to determine an accurate end point **[1]**; measure the temperature at more regular intervals or use a temperature probe to read the temperature constantly **[1]**.
**Or** **Any other valid inaccuracy and improvement.**
**11.4** Hot water may causing scalding if spilled **[1]**; take care when pouring hot water into test tubes and beakers / ensure the water has cooled completely before moving the test tubes or beakers **[1]**.
**12** A correctly drawn diagram showing three accurately drawn ray lines **[3]**; and a correctly drawn arrow to represent the image **[1]**

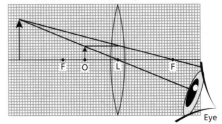

**13.1** **From top to bottom:** 1500m, $3 \times 10^{-8}$m, $5 \times 10^{-7}$m, $1 \times 10^{-11}$m **[3]** (2 marks for two in the correct position; 1 mark for one correct)

**13.2** wave speed = frequency × wavelength / $v = f\lambda$ **[1]**; $f = \dfrac{3 \times 10^8}{300}$ **[1]**; $= 1 \times 10^6$Hz **[1]**

**13.3** acceleration = $\dfrac{\text{change in velocity}}{\text{time}}$ / $a = \dfrac{\Delta v}{t}$ **[1]**; $a = \dfrac{10}{4}$ **[1]**; $= 2.5$m/s$^2$ **[1]**

**13.4** 10m/s **[1]**

**13.5** distance travelled = speed × time / $s = vt$ **[1]**; $s = 10 \times 4 = 40$m **[1]**

**13.6** resultant force = mass × acceleration / $F = ma$, $F = 1.5 \times 25$ **[1]**; $= 37.5$N **[1]**

**14.1** force $= \dfrac{\text{change in momentum}}{\text{time taken}}$ / $F = \dfrac{m\Delta v}{t}$ **[1]**; change in momentum = $4000 \times 0.004 = 16$kg m/s **[1]**

**14.2** momentum = mass × velocity / $p = mv$ **[1]**; $16 = 0.05 \times$ velocity, velocity $= \dfrac{16}{0.05} = 320$m/s **[1]**

Remember to change the mass from grams to kilograms before carrying out the calculation.

# Notes

# Notes

# Acknowledgements

The author and publisher are grateful to the copyright holders for permission to use quoted materials and images.

Cover and P1, Jurik Peter/Shutterstock.com; cover and P1, Emilio Segre Visual Archives/American Institute Of Physics/ Science Photo Library

Published by Collins
An imprint of HarperCollinsPublishers Ltd
1 London Bridge Street
London SE1 9GF

© HarperCollinsPublishers Limited 2016

ISBN 9780008326722

Content first published 2016
This edition published 2018

10 9 8 7 6 5 4 3 2 1

British Library Cataloguing in Publication Data.

A CIP record of this book is available from the British Library.

Commissioning Editor: Emily Linnett and Fiona Burns
Author: Nathan Goodman
Project Manager: Rebecca Skinner
Project Editor: Hannah Dove
Designers: Sarah Duxbury and Paul Oates
Copy-editor: Rebecca Skinner
Technical Readers: Colin Porter and Peter Batty
Proofreader: Peter Batty
Typesetting and artwork: Jouve India Private Limited
Production: Lyndsey Rogers
Printed in the UK by Martins the Printers